C
EMP

8/83

AF772529

# THE HAPPY HIGHWAYS

Edith Inglesby

In this very elegant book of nostalgia a gracious lady looks back through rose-colored mists to her childhood summers in North Carolina. With her we smell the fragrant meadows, love the beautiful horses, relive the days of youth, with its pains and promise. It was a way of life that is gone, but here we are able to recapture it at its best.

L PECK

EDITH INGLESBY

# THE HAPPY HIGHWAYS

Illustrated by Laura Peck

John Curley & Associates, Inc.
South Yarmouth, Ma.

**Library of Congress Cataloging in Publication Data**

Inglesby, Edith.
The happy highways.

1. Large type books. I. Title.
[PS3559.N396H3 1983] 813′.54 82–25265
ISBN 0–89340–574–4



**Published in Large Print by arrangement with Tradd Street Press.**

**Distributed in the U.K. and Commonwealth by Magna Print Books.**

*Printed in Great Britain*

To the Memory
of
Mother and Aunt Florence
and
Flo and John

# THE HAPPY HIGHWAYS

# *Journey of Endurance*

To remember Highlands just before the first World War is to think of a different place. Nothing remains the same; but the mountain village I knew as a child has

> "Melted into air, into thin air . . .
> The cloud-capped towers . . . dissolve
> And . . . leave not a rack behind."

And if I am to recall those halcyon days, I must start at the beginning which was the momentous affair of the journey from Savannah, Charleston, and New Orleans. A good many Englishmen similarly employed lived in those cities then. I think they loved Highlands as did my English mother because it reminded them of home . . . cold, wet and beautiful. Like stout Cortes and his men upon a peak in Darien they came, look'd at each other with a wild surmise . . . and returned each summer until they went home to war and, in many cases, death.

Cotton men had three and a half months leisure when cotton was not "moving." But the wives and children stayed for four months. The men came and left and returned. It must have taken an appreciable time for parents to recover from the journey. Ours began in a Pullman berth enroute to Augusta, Georgia. Much later it occurred to me that was a short ride for overnight and Mother confessed the train spent most of the night on a siding. Arrived at Augusta we breakfasted at the station. Once our nurse asked to "just run 'round the corner" to visit some relations. The conductor must have been a humane man for, seeing Father's agony when she did not return, he held the train for her. Time was not important. We crossed the Savannah River and slowly, in a mass of smoke and cinders, meandered through upper South Carolina.

The trip from Augusta to the foothills was made in a series of day coaches. The difficulty for Mother was that there were no dining cars and a lunch would have had to be prepared in Savannah that would last without ice for a night and a day. Hard-boiled eggs fitted this and had the merit of being a rarity as Mother considered them poison to the young. I simply adored hard boiled eggs but, alas, this for me was only an early and grim lesson in the immutability of fate. But I never learned, year after year, smoke, smell and swaying train having done their worst, I was pea green with car sickness. To this misery was added that of watching my cheerful sister guzzle my favorite food at the one time of year it was available and at the one time of year when food was abhorrent to me. Usually I was a notable trencherman.

The servants went in one coach, we in another, and Father took refuge in the smoking car. But he did not get off lightly. In those days women were throught flighty things so he had charge of all the tickets . . . ours, the servants', the two Charleston great-aunts', and once those of the wife and children of a friend who wisely was "coming up later." All this was too much for an onlooker in the smoking car. As Father produced tickets for his multitude in two coaches, bug-eyed, he blurted out, "For God's sake, man, are you a Mormon . . . or what?"

But "even the longest river, etc." so in the late afternoon, and in the final day coach, we arrived at our destination . . . for the second night. It was Baird's Hotel in Walhalla, South Carolina. Dining out was a rare treat for me and, having recovered from my car sickness, I still remember the wonderful fried chicken and the terrible green tea.

Next morning began the nicest part of the journey . . . in the childrens' eyes. The road from Walhalla to Highlands went straight up over unpaved mountain roads and we did this exciting forty miles in a hack drawn by two horses. A hack was a springed conveyance with shiny leather seats . . . three of them, on each of which three people could sit. The heavy luggage went in a covered wagon and it was with this in mind that Mother had to pack suitcases to last any emergency. Once the driver, having taken too much "moonshine," had tipped over the wagon and all the trunks, and the "bale" and the golf bags rolled out and down the mountain until stopped by trees. It was a week before they caught up with us. The great-aunts never forgot this and fluttered anxiously about Father beseeching him to ascertain if the drivers "had been drinking," a Charleston gentility for being drunk. Father was more concerned with the conditions of the roads. "Has there been much rain?" he would inquire. As well he might. Rain washed deep ruts which made it impossibly slippery in the slimy wet red clay or revealed slick sides of boulders over which the hacks slid and listed precariously toward a drop of one thousand feet. We would all lean in the opposite direction to help balance. Pine Mountain was so steep we got out to save the horses.

We stopped for lunch at a farmhouse called Russell's where the wonder of the food was matched by the thrill of knowing "Mr. Russell hung himself in the barn." He had nineteen children . . . and at this point Father must have had a sneaking sympathy for him. One year a hack broke down and we had to spend a night at the farm, our third enroute!

To stretch our legs before resuming the ride our nurse would take us on a walk, or we would take her on one, we were far more eager than she was. Here, near Russell's, we knew that our Savannah River had its beginning. The Army Corps of Engineers has designated the confluence of the Seneca and the Tugaloo much farther down stream as its place of nascence, but for me the River will always begin at

Ellicot's Rock that huge boulder flung up by the Chattooga River, the rock marked Latitude 35, which history has singled out as the survey point where North Carolina, Georgia and South Carolina come together. The Chattooga has a long way to go around the top of South Carolina before it joins the Toturo to become the Tugaloo, but from that point it goes straight down the map between South Carolina and Georgia. To us the little boisterous stream seemed very remote from our great red river at Savannah.

Summer after summer in Highlands Father "took a place" which belonged to the Misses Ravenel of Philadelphia. They were a branch of the Charleston family which owned large tracts of land in Highlands. The place Father rented was named Islington ... Islington was named after an old part of Charleston which carried the name of a fashionable district of eighteenth century London. We put up at a boarding house until Mother thought Islington was in a state ready to receive us, and the trunks had arrived ... to say nothing of the "the bale." The bale was a huge square of brown canvas into which were thrown warm coats for September, riding clothes, toys, books and what not. This great bundle was then sewed up with twine held by a great needle known as "the bale needle," the locating of which each year began the travail of our odyssey.

In time, the second night enroute was eliminated; but we still began the journey in a Pullman car bed. This one took us to Atlanta where we boarded a day coach for Dillard, Georgia. Here, instead of a hack, we were taken up to Highlands in Ford cars whose drivers stopped at every stream to replenish the radiators from each of which the tops had long since been blown off. In a cloud of steam we negotiated rocks and holes and forded (no pun intended) streams. It was a lot quicker but for us children not nearly as much fun. We liked the horses. Early on we learned their names and sexes and by the time the forty miles was done, they were our dear familiars. It was fortunate cotton men could take such holidays. A journey like this was, as the Prayer Book says of marriage, "not entered into unadvisedly or lightly."

As much as Father loved Highlands, its remoteness, its good society, its altitude (Satulah Mountain is 4543 feet above sea level) as much as he enjoyed the good fishing, the hiking trails and waterfalls, he would not have countenanced any place without a golf links.

They were links then, not courses. With the aid of mountaineers and college boys he built the first two in Highlands . . . long before Bobby Jones was ever heard of. I don't remember the inception of the first; but I do remember the excitement attendant upon the birth of the second.

It was at the present, and quite different, golf club in Highlands that I got a shock. A young girl near me called out to a contemporary that she was "going to run down to Atlanta to see a movie." Nothing, but *nothing*, could have brought home more sharply the changes time had wrought.

# *Arrival*

We had, oh blessed thought, arrived at Highlands! The long summer stretched before me in a haze of delights. Such was my content that subconsciously I was thinking: Nothing can ever go wrong. We are in *Highlands*!

It was not a pretty village; not even a picturesque one as far as its Main Street went. One had to know where all those rhododendron-lined driveways led. We will, in time, go to Faraway.

Now that Highlands has, unfortunately, been discovered, Faraway may not be the grandest estate. But it always will rank first with me in splendor. Islington, where we lived, was bigger, white and surrounded, upstairs and down, by pillared piazzas. Furthermore, and this was a large plus in my eyes, it had the only mounting block anywhere around. But compared with Faraway with its oriental rugs, its Steinway grand piano and its long low, leaded-pane windows in drawing and dining rooms, and its wonderful barn with Kentucky thoroughbred horses, Islington was just a big, comfortable summer country house.

Highfield, where Flo and John Perry summered when not in England, (they were from Charleston) had lovely distant views and a most enchanting "rhododendron path". This was entered through a rustic gate on the Walhalla road and wound its flag-stoned way up to the piazza. Piazza is a Low Country word for porch or gallery; our use of it never failed to amuse Mother, and I must admit that most houses in Highlands had porches. Porches, galleries, verandahs, piazzas – we used 'em all, but the Perrys were from Charleston, they said *piazza*. Sitting there on the piazza one would hear steps. Who could it be? Then at last someone would emerge out of the dusky rhododendron tunnel into the light. It was often John, sent back to the farmer for more milk. Sent back on foot although John disliked walking. The quart mason jars the milk came in were difficult to handle on horseback. Once he tried to invent milk bags to hang from the saddle of his horse. This, however, was not a success.

Perhaps the nicest thing about Highlands

was the people it brought to it . . . the "summer people." This is not to decry the mountaineers, many of whom we knew and loved. Some of them – the girls – were beautiful. They were pure Anglo-Saxon with the bluest eyes and the sort of wheat-colored hair that goes with them. And, seemingly pure largesse, it often was curly.

We had friends from Charleston and New Orleans whom we greeted each summer, compared inches of growth, played tennis and rode. But it was being with Flo and John Perry that made that endless journey worthwhile.

Mother and Father had reunions, too. If Father fished with Mr. Eskrigge and others, Mother gave and received crumpets and countless cups of tea from their waves. We were a very English community. It was a lovely, leisured life and, looking back, indeed a "land of lost content."

"The Atlanta People" at that time seldom owned houses but stayed at The Hall House. They dressed very smartly and illuminated parties when invited. But by and large the "Atlanta crowd" stayed to themselves.

If Main Street was ugly, and it was, one should except the end where the Episcopal

Church of the Incarnation stood. One of the "Elliott boys" was Huger Elliott who was an artist and had something to do with the Metropolitan Museum of Art in New York. To embellish the chancel he painted figures from the Bible using his family as models. It fascinated us and in whispers that very much annoyed Flo's governess, Miss Jones, we would identify them. Old Dr. Elliott was Moses and the young Saul in armour was Percy Elliott who *was* young, though we could not very well imagine him so accoutred.

The greatest reunion of all, however, at least to me, would be with the horses, Moonlight and Magnolia and the horse that pulled the dogcart, Caesar. We did not own them, they belonged to a man called Otto Russell, (he whose father, overcome by the arrival of his nineteenth child, hung himself) and were kept in winter at a town near Walhalla, South Carolina. I knew Moonlight and Magnolia should be *at* Walhalla. Surely these two horses belonged in the abode of the gods.

Correspondence between Father and Mr. Russell began in May and soon after we arrived, but never soon enough for me, there they would be – the treasures of my heart. At least so Magnolia was. Moonlight did not arouse in me the transports my dear "Maggie" did. But as we will see in a future encounter with the spavined "Ralph" all horses held my regard, and Moonlight certainly was next to my cherished "Maggie." One of the great-aunts from Charleston who was hearing my prayers, objected to the inclusion of Moonlight and Magnolia; these were, she insisted just dumb beasts. This was too much for me. (Had she but known it, she was interrupted just before mention of Flora Macdonald, Caesar and a dish-faced horse I had recently encountered, "Lily Langtree," rented that summer by my contemporary, Claude Northrop, from Washington, D.C. but whose mother, as he was quick to let one know, was a Grimke from Charleston.) To retrace that wandering thought ... the "just dumb beasts" enraged me so, I hit my aged relative a shrewd blow on the shoulder. To her credit she did not tell Mother.

Obsessed with horses as I was, Highlands was a most glorious place. I did gather that to other people it had other charms. First we had arrived, next we would have possession of Moonlight and

Magnolia. We would have horses in the barn. I loved the barn ... the smell of horse, the scent of new hay and the smell of leather softened by neat's-foot oil. I liked to go up in the loft and fork down the hay ... for bedding and for food. And there were oats, too. Liking the warm, mushy altogether delightful Quaker oatmeal I had for breakfast, I once tasted the horses' oats. A mistake.

I was not really insensitive to the setting of the small village perched on a bit of Satulah Mountain. We got there in time to see the kalmia, or mountain laurel, *ivy*, the mountain people call it, spread its fairy-like bloom under the tall forest trees, disappearing up the steep slopes in clouds of pale pink. Some years mountain after mountain would be as exciting as an enchanted garden. Occasionally we arrived in time to see the flame azalea, but only on the heights. After the laurel would come the rhododendron, great pale clusters of flowers on heavy dark leaves supported by twisted stems like Arthur Rackham drawings. Mother kept two crockery jars filled with rhododendron on each side at the top of the wide front steps. Inside, the house was filled all summer long with wild flowers. Wild flowers don't last, but there were always plenty more, that is until the tall grass holding the brown-eyed Susans, the daisies, the clover and Queen Anne's lace were all ruthlessly scythed down.

And everywhere one looked were "views." Not that we called them that. They were just the pleasantness that surrounded us to which we were so accustomed that it lay deep inside, this consciousness of bounty, not mentioned. To our left was Satulah, behind us was a low range of mountains that at one place had a great flat rock side. This was known as Sunset Rocks. Looking beyond the village was Bear Pen and Little Bear Pen.

There were all sorts of falls. The small ones that splashed down the rocks of little rivers and the large majestic ones that, with a great sound of rushing water, dropped from a tremendous height to join the rocky stream far below. Cullasaja Falls was the most exciting ... and the most beautiful. But when I was very small it was a little frightening, too. Now proper steps have been built down to where one may stop and view it all. But then, Mother and Father, each clutching a child firmly by its hand, would climb precariously down the steep, slippery incline, hazardous with rhododendron roots to walk dry shod under the Falls when, triumphantly, they reached the bottom. It was splendid but frankly terrifying. The sound of the Falls was deafening and I went only to boast of having gone. The backbone of most youthful conversation is boasting, pure and simple.

From the piazza at Highfield one could see Billy Cabin mountain and the snowy cliffs of Whitesides. This latter was a wonderful mountain with a wonderful story. A few mountaineers were making merry with "moonshine." One leant a little too far over the cliff . . . and fell. His progress to eternity, however, was stopped by a smallish tree. A rope was tied around a man who was lowered, grabbed the dangling one, and together they were hauled back to safety. For this feat . . . achieved, Father said, only because they were too drunk to know fear, the hero was awarded the Carnegie Medal. So much for Dutch courage.

And there was Yellow Mountain around whose top, one hears, progress has manifested itself in a ring of condominiums. But let us forget this horrid intrustion. This is a tale of Highlands just before World War I, of

> "Daisies pied and violets blue,
> And lady-smocks all silver white,
> And cookoo-buds of yellow-hue."

And, of course, horses!

How to describe Moonlight and Magnolia? Only one thought springs to mind . . . Pegasus. But he had wings and though our dear companions had quick hoofs they had not wings. They only seemed, at times, to have them. We could hardly wait to see them, but Mother said one thing at a time. This couldn't come until we had the house in order and the barn ready for them.

They were bay fillies of about fifteen hands with shiny, rounded quarters and slender, pretty legs. Magnolia had benevolent eyes and a compassionate disposition. If her rider was inexpert and given to losing the stirrups, she would take the bit firmly in her teeth and lead him gently on a safe if unexciting ride. Moonlight showed no such compassion. She would dance impatiently, her eyes flickering nervously, showing their whites.

Sometimes her coat would be black in patches of sweat. Mother said she was high strung and should be handled with care and affection . . . which Charlotte did. It was for this reason she never let anyone but John and me ride Moonlight. And as we had our own horses Charlotte and Moonlight pursued their solitary search for wild flowers. Charlotte, one arm full of flowers, the other holding Moonlight's reins, would move slowly over the meadows. Beside her Moonlight picked a careful way, her eyes looking about a bit apprehensively. Then there was also Caesar who pulled the dog cart. Caesar was fat, and piebald. We called him Mother's horse and condescended to him. Mother who used to ride only the most spirited of horses and sidesaddle at that, had given up riding when she discovered that Father didn't enjoy it because of a back injury and preferred golf. She had taken up golf. And it was she who had given Caesar his name, as being totally inappropriate.

While I waited the arrival of the horses I renewed my knowledge of every nook and corner of Islington's grounds. I went to the very back and climbed some apple trees from which, in the autumn just before leaving, we sent home barrels of red-cheeked fruit. Once an elderly great aunt, supervising the packing of the barrels, lost her spectacles. When she was unpacking the apples that winter in Savannah here were the missing glasses ... lying unbroken among the fruit. I investigated the raspberry canes from whose fruit this same great aunt made such delectable jelly-rolls for tea. I climbed pine trees, the *pinus Elliotti*, and fell out of a few of them. But as my fall was slowed by every bough, I was pretty well intact when I reached the ground.

It was a time of waiting, but it had its compensations. Each morning a pale sun turned hot and yellow at noon to bake down into the roots of the grass. I would burrow into this aromatic fastness, so tall and concealing. The smell of hot grasses would be mixed with the fragrance of wild strawberries. Occasionally I would reach out and pop a berry into my mouth. At intervals an ant would scuttle across my face. There were bees, murmurous and soporific. I loved it. It was a secret place. Not Savannah, not Highlands. A place where grasses and flowers were larger than life and a little unreal. I would lie for hours, thinking. Here in the scent and sound of summer I thought of horses. Horses were the real business of summer.

And then they arrived! Crunkleton, "our man," was a tall, taciturn mountaineer who walked three miles each day up from his valley. He seemed to endure rather than like us. But he took excellent care of the horses. This morning he was fuming because he had asked us to rub up the tack with neats-foot oil. And we had not done it. "I muck out the stalls and go down for their feed. You-uns ain't do nothin' but ride these here horses."

But I didn't mind his grumbling. I was caressing Magnolia with ardor. Then I sneezed. "What did I tell you?" said Crunkleton. "You got to curry that beast, she's full of dried mud from rollin' and she needs clipping, too . . . still got all that winter coat."

It was all quite true. And we set about restoring what I fondly – if mistakenly – thought of as "a show condition."

Then one morning happiness was complete when Crunkleton said, "Them Perrys is here." There had been a letter from Flo but the date of arrival was vague. Flo and John, in age, were just between

Charlotte and me. Charlotte, I looked up to as a superior but Flo and John both seemed just my age. This fact I foolishly made known to John who let me know in no certain terms that I was his junior by five whole months.

"Flo and John are here," I called. There was a rush of feet and Charlotte joined us. "Why didn't you let me know?"

"Flo and John are here!" We danced around and around the barn.

Charlotte found Moonlight's saddle blanket and turned to Crunkleton. Grudgingly he settled it and put the saddle on. I wondered why she didn't saddle Moonlight herself. We usually did.

As Crunkleton tightened the girth I said, "I wonder if they've grown much. John's been to boarding school – in England."

"Being at boarding school wouldn't make him grow. Just make him more bossy," said Charlotte as Crunkleton gave her leg up.

I admired John. He rode well. But he was bossy enough already.

"I wonder if Aunt Florence will think we've come too early?"

"She won't know. It's that Miss Jones who won't like it. She never likes anything.

Probably got Flo doing something ... irregular French verbs most likely." We could think of nothing direr.

I shuddered. "Well, let's go anyway."

"Aunt" Florence was aunt by affection. Uncle Hampton was "uncle" because he was her husband. He produced in us rather the feeling of a vacuum. He was *there*, and always very well dressed. But we never really saw him.

We were all here ... the most important ones. Mother, Father, Charlotte and me. And "The Perrys," ... Aunt Florence, Uncle Hampton,F Flo and John, and the horses! The summer had started.

# *Finesse*

John and I had been down at the paddock practicing jumps for the gymkhana. Flo and Charlotte ... not uncriticial spectators ... came back to the verandah with us where Mother and Father waited for lunch. At this point Miss Jones arrived. Miss Jones was Welsh, fiery and impatient. She was Flo's governess, but during the summer months when he was home from school she also conducted a running feud with John. This, though, was known as John's holiday reading.

We watched them depart down the drive, the two children mooching along, Miss Jones walking with a brisk air, an offended tilt to her head.

Mother looked at me and sighed. "I think you hurt Miss Jones' feelings." I was completely astonished.

"It's quite time you learned what *not* to say. When Miss Jones told us about her father and the Eisteddfod ..."

"The Bard of Llangollen," Father

murmured.

Mother's annoyance now included us both, "It's all very well, George, but Edith should not have said 'Yes' when Miss Jones asked if she'd told us about him before."

"But she *had*," I complained, "several times. I couldn't tell a story, could I? Could I?" I repeated.

"You might have . . ." Mother stopped, exasperated.

"Tempered truth with discretion." Father jingled the coins in his pocket, a way he had when amused. "You-er-finesse it."

They had just come in from golf and Father was having a whiskey and soda. It was July, but in the North Carolina mountains he wore woolen plus fours, woolen socks and a viyella shirt. Even so he looked cool and distinguished. Mother, her lovely English skin glowing with color, had her dark hair gathered back in a figure eight and her linen skirt stopped daringly at her ankles.

"Miss Jones," I began. But Father, bored with it all, finished his drink and set down his glass. "Please tell Legacy to serve lunch."

"I wish you would not call Haskell

'Legacy,'" Mother objected.

"Legacy. And sole!" he added bitterly.

Haskell had come to us from Uncle Peyre in Charleston. Together they had had an establishment that was the envy of that city. Uncle Peyre, tall, urbane, adept in the art of living, Haskell, black, rotund, the perfect cook, valet and friend. By aid of what Father called "some fancy side-stepping" they had both, in their respective spheres, eluded matrimony. Uncle Peyre's rather worn handsomeness, his charm and courtly air was the patina on Haskell's wizardry with food and drink. Their parties were famous. However, death . . . Uncle Peyre's . . . interrupted the perfection of this menage.

All Charleston thought Uncle Peyre, "in sea island cotton," was rich and that Father, his only relative, would be his heir. Father shared these views. Mistakenly, it appeared. What Uncle Peyre left was debts. As Father put it sadly, Uncle Peyre's lack of interest in sea island cotton was exceeded only by his optimism on the stock market.

A Gilbert Stuart of their common ancestor, plus some fine old furniture, sent Uncle Peyre debt-free to heaven. Haskell, Father triumphantly brought back to

Savannah.

Mother, hardly able to believe her luck, was determined to keep it. Haskell, she admonished us, was not used to children. We must not tag along after him or bother him in the least. Haskell, however, was accustomed to livelier company than Father's house afforded. He and Charlotte and I were entirely compatible.

At lunch Charlotte announced that we were going to see the Staubs that afternoon. We were eating the mountain trout that Father had caught.

"Professor Staub," Charlotte said rather grandly, "eats jellied eels."

"How on earth do you know?"

"Charlotte and the Professor talk and talk," I exclaimed with pride, "they talk about Halley's Comet."

"A subject I should think soon exhausted," said Father. But Charlotte, assuming the look of a scientist, said, "We talk lots about Halley's Comet. The Professor has seen it twice and I saw it once."

"Edith saw it too."

"I can't remember," I mourned.

"Oh," cried Mother, disappointed, "I woke you and took you to the window. So you could tell your grandchildren you had seen Halley's Comet."

"Well I can't." I was beginning to regret my honesty. Certainly I had information enough from Charlotte to regale any number of descendants.

"They will have to take it on faith," Father offered.

"Anyway," Mother said brightly, "be sure to talk to the Staubs in French. It is a wonderful opportunity."

"Swiss French?" asked Father.

"Well – I," began Mother.

"Why don't you take them on yourself?"

"She does," I affirmed, "She's always saying 'Pas devant les domestiques'."

"A necessary phrase. But it's no good trying; I speak French with the strongest British accent. Though, if it comes to that, that woman who teaches them French at their school ... she speaks *English* with ... what is it you talk in Savannah and Charleston?"

Father gave a hasty glance round the dining room. Haskell had gone to his pantry. "The *Nigrahs*," he said rather stiffly, "speak Gullah. I presume that's what you mean. Actually," he admitted, "we do get a tinge of it in our accents."

"That's it . . . Gullah. Well, if it colors Miss What-you-may-call-it's English, I should think it must color her French too. You would think that Florence would talk to them some. She has a wonderfully good accent. But she never sees the children except at a nice distance."

"Aunt" Florence . . . aunt by affection . . . was from Shopshire and Mother was from Surrey. In England, acquiantances, they had, by what Aunt Florence termed "the most extraordinary circumstances" both married Charlestonians. Aunt Florence had met Uncle Hampton when he arrived at England on his Grand Tour. Mother met Father in Savannah where he had been sent by his Charleston cotton firm. Mother was visiting a great-aunt who had taken an odd fancy to move from London to Savannah. "She was an eccentric," Mother would say. "I'd been warned no good would come of it." But she was only teasing.

Meeting again at Highlands, Aunt Florence and Mother had revived that long-ago acquaintance into the close bond that now enveloped us all. Father, a cotton factor, had long holidays when cotton was not "moving." Uncle Hampton enjoyed twelve months leisure thanks to Aunt Florence's income and a lethargy native to his being. The two families, we at Islington and they at Highfield (nostalgically named for Aunt Florence's home in Shopshire) were almost like one. Too much so, I ometimes reflected, until the relentless surveillance of Miss Jones.

Getting back to the Staubs I said, "I talk to Miss Odile while Charlotte and the Professor talk astronomy."

"What do *you* talk about?" Father asked politely.

"Oh," I said vaguely, "things. She showed me some clothes once."

"Clothes!" Mother's incredulous glance took in my bedraggled dress and matching bloomers, equally abused. Bloomers, instead of lace trimmed panties were an innovation for little girls. Dresses although short covered them decently. My legs were bare, except for some droopy socks and open sandals. For the gymkhana I would wear my nice habit and boots, but I couldn't wear such fancy clothes every day. I had washed my hands for lunch and had hastily brushed the strands of curly hair that had escaped from my pig-tails back out of my eyes. "It is not a subject that I should have thought would appeal to you."

"It's all her things I like, really. She has a clock with a knight on horseback who comes out, you know . . . and he rides up and down when the clocks strikes. Then he goes back in again." I was still pondering the wonder of it. "And," I exclaimed, "she went to a horse show at St. Cyr. But she didn't seem to know much about it." My tone indicated that this time Miss Odile had let me down. "I wanted to know if they rode the forward seat . . . we learned about St. Cyr in riding lessons. But she just said about what she wore. That was the clothes I was talking about. She made me look at this dress."

"I don't suppose you liked that much."

"No. But I looked."

"A concession."

"The bronze shoes that went with the dress were real pretty," I said to give the devil his due.

"Florence," went on Mother whose abstracted look showed she had not been listening, "got this Miss Jones partly because she's good at French. She studied at Tours, I believe. *She* talks to the children."

Charlotte and I exchanged pitying

glances. We knew Miss Jones never spoke a word of French she could avoid. The truth was Miss Jones hated France. Recently we had speculated about this.

"Perhaps she had a love affair there," Charlotte suggested. She was beginning to take a romantic view of life.

"Miss *Jones*! Who'd look a her? Not even a Frenchman," John's scorn included them both. "Maybe they made her eat frogs," he added, "they do eat 'em in France. That's why the French are called 'Frogs.'"

But I knew better. France was a place where the inhabitants drank light wines and danced continually. "Sur le pont d'Avignon." Besides my georgraphy book said so.

"If you're going to the Staubs," said Mother, "you can carry a note for me. It will save postage." Mother was fond of little economies; large ones did not appeal. Postage was two cents at this point.

"It's an invitation to the tea on Thursday."

Father received this information with ajpained look. We sympathized. An afternoon party . . . or rather its comings and goings which were all we saw of it . . . we found dull. An evening affair was different. At one of these we would hang over the banisters of the upstairs verandah watching the guests arrive. At the gate first the flickering carriage lights, then, at the open front door, carriage and alighting occupants took on a romantic and unfamiliar look in the sudden illumination against the night. The guests on foot arrived by the path carrying lanterns, the light bobbing up and down as they walked. Once on the verandah they would blow out the flame and set their lanterns down in a corner, each tagged with a label bearing its owner's name.

We would tear ourselves away from this delightful scene to rush indoors. Here we would cling to the stairwell rail, looking down at the so-to-speak finished product ... guests now divested of cloaks and wraps and often umbrellas. (It rained frequently at Highlands.) They would come out of the library, through the hall and into the radiance of the drawing room. An *afternoon* party was tame by comparison.

After lunch we set out by the path for the Staub's house. With us was our dog, a golden cocker sent to Mother from Scotland one summer. Father said the connection of Scotland, North Carolina and the feminine sex made the name, Flora Macdonald, unavoidable. Flora was getting on in years now and she was much too fat. She was not a tidy dog; burrs stuck to her, leaves nestled in her luxuriant feathers. She did not care, nor did we. Mother did, and demanded a certain ritual brushing and combing, but the results weren't spectacular.

We, however, were spruced up for the occasion. Charlotte had on a dress with roses embroidered on the "bertha." I was ... for the moment ... immaculate in white linen. "It's no good my trying to put her in anything else," Mother once exclaimed in my hearing. "It's good Irish linen ... from Hogan's in Savannah. The only thing that can take the boiling Edith's clothes have to have. Currying the horses, nearly driving Crunkleton mad helping him soap the harness, falling in the pond and coming out duckweed all over ... Really, linen's the most practical thing in the end." This was before the days of blue jeans.

Skirting the village we took the road that led to the Staubs. We made slow progress. Flora conducted endless research among the rhododendrons and Charlotte made sorties gathering wild flowers for her friend the Professor. She ran about picking Indian paintbrushes, butter-and-eggs and the sweet scented bergamot. Holding them in both hands they fanned out in a circle of color against her dress, but she kept Mother's note separate and clean between the thumb and fingers of her left hand.

Then we came to my favorite place . . . Harbison's Brook. The water came tumbling out escaping through the fence of Harbison's meadow to flood in sparkling torrents under a hollow-sounding bridge. It plumed up in white sprays over the rocks, and then made brown pools in quiet places. To me it was Tom's brook, in *The Water Babies*. Tom's brook before he went to sea. I almost believed Tom still lived here; at any rate I always looked for him, under the bridge. Charlotte discouraged me now. "You'll get soaking wet, you always do." But I insisted on trying to ford the stream stepping gingerly on the flat stones and succeeded only because Charlotte, scolding, pulled me across the last stone and up the bank. I didn't have time to look under the bridge. We left Harbison's Brook and resumed our journey. We did notice that Flora Macdonald had not been as careful as we were in crossing the brook. She was muddy and her wetting made her look skinny. She didn't look like herself at all.

Brooks enchanted Aunt Florence. The one at Highfield dropped sharply away on the lower bank. She dubbed it "Becher's Brook." "Like the one at the Grand National," she'd say, beaming. Mother had never seen the Grand National. She did not care for racing, although she too enjoyed the sparkling streams. Uncle Hampton, who was contemplative and scholarly, liked to ponder this essential difference between them. It was a difference that in no way lessened their affection At names such as Whistler or Augustus John, Mother's face would light up in happy recollection. Grandfather had been a portrait painter, a liberal and a friend of Gladstone's. Aunt Florence's father did not know Gladstone; nor would he, one felt, have cared to meet him.

Carrying Mother's note, we three ... Flora Macdonald, disgustingly wet and muddy ... followed the road that led to Sunset Rocks. Near the top of this long, high plateau was the Staub's house. It was set on a piece of flat land that looked niched into the side of the mountain. From it one could look up to the top of Sunset Rocks and down to Harbison's meadow with the village beyond.

Professor Staub, a Swiss astronomer, was reported in the village to have taught in some unnamed Mid-Western university. Ninety-two years of age, he lived now, summer and winter, at Highlands, with his daughter, the rather fluttery Miss Odile. Why they had chosen it? How indeed they had heard of this remote mountain hamlet in the Southern mountains was a puzzle. As a matter of fact, there was quite a cosmopolitan flavor to our village. Mother used to say, "How did all this conglomeration of oddities, including myself, arrive here? It's amazing!"

The Staub's garden was beautiful and it smelled divinely of pinks. Everywhere was color and bloom. "It is like England," Aunt Florence would say. As she made this observation of anything that took her fancy, we quite rightly discounted it.

The Professor sat wrapped in a steamer rug in a garden chair placed in the sun. "Ah," he cried in delight. "Voici les enfants." He beamed and, encumbered with his rug as he was, adumbrated a polished bow. “Odile, viens ici,” he called. “Les petites Inglesby sont arrivées. Les jolies petities,” he added smiling upon us. And Charlotte, so flown with praise, went forward and presented him his bouquet of wild flowers with a curtsy. Miss Jones encouraged this politeness. Flo and I, solid and self-conscious, hated it, but Charlotte, slender and graceful, enjoyed it. Curtsying became her.

Miss Odile came flying out. she advanced in swift, darting movements like a swallow. Birdlike in other ways, she had a magpie’s acquisitiveness. All her treasures from the past she kept hidden away. She would tell me of events in which they figured and bring them out before my bedazzled eyes. The scent of mothballs became for me the odor of romance. She was Scheherazade; I, her enchanted listener.

We presented her with Mother’s invitation. During my brief encounter with Harbison’s Brook, when Charlotte had to pull me up the bank, it had been carefully placed on the bridge railing. Miss Odile was delighted at the thought of a party. “It will be like Geneva.” Fortunately we were in no position to disillusion her.

The Professor wished to return to the house where Charlotte and he would be near their maps and celestial globes. But Miss Odile urged him to tranquilize himself and remain in the so healthful sun. I was amazed. At ninety-two was one still being bossed? But Miss Odile was seventy.

The Professor won out. Miss Odile took his arm. Charlotte carried his rug and Flora Macdonald and I followed the procession. At the door Miss Odile stopped. She looked doubtfully at Flora; then with a little sign invited her in. "La jolie petite favorite," she said kindly. We loved Flora devotedly; but it would not have occurred to us to call her pretty.

Charlotte and the Professor obviously yearned to get to their studies, and I pined to hear Miss Odile's stories. But the ritual must be observed; we must partake of cakes and wine. Charlotte had confided to me that more than anything she hated the taste of the Staub's seed cake, yet here she was sipping the unaccustomed, if watered, wine, eating the despised cakes with apparent pleasure. I did admire Charlotte's manners.

Afterwards Charlotte and the Professor retired to some scholarly fastness and I prepared to be fascinated by Miss Odile. My hostess, however, was not her usual self. Quite plainly she was thinking of something else. When, I asked myself, would Miss Odile remember to bring out her wonderful things. But Miss Odile seemed almost oblivious of my presence.

Then she turned to me. "Do you remember the dress I showed you, the one I wore to the horse show at St. Cyr?" This was more like it. Prepared this time tojpin Miss Odile down to facts as to the forward-seat I nodded vigorously.

"That dress I shall wear to your mama's reception. And," with excitement, "bien entendu, I shall wear with it the bronze slippers."

"They are beautiful."

"They set off the dress," she agreed not without satisfaction.

Actually I did admire her slippers. Miss Odile's feet were small and quick and they set themselves down so surely, so deftly ... almost as if they thought for themselves. Such clever feet. I should like to see them in the bronze slippers.

The morning of Mother's tea Charlotte went to spend the day with a friend to whom she was united by a common bond of wild flowers; formerly it was paper dolls. After a ride with John I came to the house in search of Haskell. I found him polishing silver in the pantry. He looked gloomy. Stopping in his rubbing of the great silver tea urn he turned a disgruntled look at me and exclaimed; "Me and yo' Uncle Peyre never fool with all this tea."

"What did you fool with, Haskell?"

"We had whiskey, brandy, rum, port ... every kind of wine. An', of course, champagne," he added with pride.

"Didn't you ever have tea?" The mores of Haskell and Uncle Peyre was a topic of never failing concern to me.

Haskell laughed, "Sure nuf, we used to have teas sometimes ... for old ladies and gemmans."

"Uncle Peyre was an old man himself," I remarked. I could see Haskell thought this tactless. A disapproving silence hung between us.

"What kind of teas?" I asked to change his thought.

"Well, we'd have tea in the drawing-room. Mist' Peyre say to me, 'Haskell, he'd say, some of these gemmans going want dey toddy. You set it up for them in the libry; and see does dey get it. We got to use fee-ness,' he'd say."

"Fee-ness? What's that?" – Wasn't that the same word Father had used about not down-facing grown-ups?

Haskell reflected. "I think it mean handling a sichation kin' of easy like."

"Haskell," introducing the subject I had come to discuss, "I'm going to the party this afternoon."

"Yo' ma know?"

"No."

"Oh, oh," said Haskell. "*Oh*, oh."

"Not for the whole time. Just for a little while at the end." But Haskell eyed me with suspicion. "You ain't fixin' to come in this rig?" He eyed my stable stained dress.

"Of course not. I'm not even coming to lunch like this."

"Thank Gawd for that," he observed piously. "Yo' ma tell me to wear mah linen jacket. Do seem mah swishtail coat got mo' style." He sighed. "W'y you wan' to come to dis party anyhow?" It was clear he did not think highly of it himself.

"Well," I paused. I did not know exactly how to put it. "You see Miss Odile tells me about places she's been. She's going to wear a dress this afternoon that she wore last at a horse show in France. Haskell, you know she saw the haute ecole riders. Their horses . . ."

Haskell knew my passion for horses. "Ain't goin' be no horse at *this* party."

"Of course not," I said crossly. "I know that, Haskell; but I thought it might be sort of fun to see Miss Odile in her dress and her wonderful bronze slippers that she wore to St. Cyr."

"What horse see the shoe?" Haskell laughed heartily at his own wit. Then he appeared to reflect. "You could git all dress up, den sasshay sort of nachel-like thoo the rooms . . . toads the end. Lak you jus' happen to be passin' this way."

It did not occur to me . . . nor apparently to Haskell . . . that for Mother my being dressed in my best hardly would appear natural. I thought it a very good idea.

"How will I know when it's near the end?" I was beginning to see difficulties. Haskell said he would come and tell me. But I was just to go in, speak to Miss Odile, then leave." "Thout you do it that way I ain't goin' to help."

"Oh, I don't want to stay long at the old party."

By late afternoon I was dressed and waiting. I had found my best dress and put it on. Unfortunately my good white shoes needed polish. This was overcome by wearing Charlotte's. Though she was several years my senior her shoes fitted me perfectly. I would grow to my feet in time, Father said. Like a puppy. My hair, released from pigtails, fanned out in curls. I just let it hang. Thus arrayed I went to Mother's room to see myself in her cheval mirror. Passing her dressing table I annointed myself lavishly with scent. As I waited I could see the rain that had been falling all afternoon had finally stopped.

Suddenly behind me there was a loud hiss. It was Haskell at the head of the stairs. "You better come," he shouted in a whisper. "They's gettin' restless. I think they's fixing to leave."

"How do I look, Haskell?"

"You looks all right; but you powerful loud. Jus' wait till you ma get a smell of you."

On this discouraging note I followed him down the stairs.

The large room was filled. Spoons clashed against cups and the buzz of voices was almost a roar. Unnerved, I looked about for Miss Odile. I got here just in time. With outstretched hand she was advancing upon Mother. Evidently she was making her adieux. That was the horse show dress, all right. Miss Odile, I reflected, looked real peculiar. This did not disturb me. It was Miss Odile's entertainment, not her clothes, that endeared her to me.

I went over and stood behind her as she exchanged politenesses with Mother. Then I remembered the treasured slippers. My eyes, travelling down, searched them out. I stood stock still in dismay. A look of horror must have been plain on my face; for Mother, who had glanced at me in brief speculation, now followed my agonized gaze to Miss Odile's feet. Miss Odile may have been wearing her bronze slippers, that adornment which she obviously considered the last word in refinement of apparel. I could not tell. What Mother and I were looking at in pity and in sorrow were a pair of large overshoes. Useful and serviceable in such weather, but hardly what Miss Odile meant to wear to Mother's party. My heart ached. I could imagine how Miss Odile would feel if she knew. Mother looked fixedly at me, as if to stem the forthrightness she so deplored. She need not have worried. This time I knew what not to say.

Leaving Mother, Miss Odile made her way through the crowd to the Library which was serving as cloakroom. The crowd, I supposed, was why no one had seen her rubbers. Or had they, and like Mother and me, forborne to tell her? I darted out into the hall. Haskell was standing near the door. "Oh, Haskell," I cried, nearly weeping, "No one could see Miss Odile's bronze slippers. She's had on those hideous rubbers all the time."

"Great Gawd," said Haskell simply.

Haskell had a kind heart and he loved elegance. He stood there sorrowing with me for Miss Odile. I could see her through the open door struggling against the flowing tide of women reaching for wraps, corraling umbrellas, chattering like mad. Apparently giving up in despair she came into the hall. At a little distance away she stood making futile efforts to don her raincoat. Haskell stepped forward and helped her. She seemed, for all her worldly experience, tired, old and confused. Fastening her raincoat she reached for the lowest button, looking down ... and saw the rubbers.

"Mon Dieu," she cried piteously.

"Surely I could not have worn these all . . ."

Resolutely I ran up to her. "Oh, Miss Odile. I saw you a little while ago from upstairs. You looked so nice in your horse show dress. And your bronze slippers. You've already put on your overshoes and your coat. I am so glad I got down before you left."

Miss Odile glanced back a moment uncertainly at the library. Then the small, rosy face . . . puckered in distress . . . suddenly cleared. "Yes," as if to herself. "Mais certainement. C'est ca, ce que j'ai fait. That is what I have done," she repeated. "I have put back on my rubbers. One gets confused."

Gaiety returned, and giving her elegant little bow of departure she made her way through the front door, her feet, shrouded as they were, still the knowing, clever little feet I admired.

Mother had come to the drawing-room door. She turned to go back but stopped and briefly touched my shoulder. "Good girl," she said. "Almost everybody's gone – wouldn't you like some tea?"

But I was too shattered by Miss Odile's narrow escape. I said I thought I would go upstairs. Not even the sight of Jordan almonds tempted me.

At the newel post Haskell intercepted me. He looked at me in approval. "Yo' Uncle Peyre's fee-nesse," he said.

The next morning my imagination had endowed Miss Odile with clearer perception than she had displayed. She must, I thought, have realized about the rubbers. It was a new feeling ... compassion for a grownup. I experienced the fact with surprise and some resentment that tables were turning.

"May I take some cakes to Miss Odile?"

After Mother's parties she often sent cakes and flowers to those whose age infirmities prevented their attending. She was preparing a basket for old Mrs. Ravenel of Wantoot. To reach Wantoot, Crunkleton would have to pass right by the Staubs.

"The Professor never got any at all," said Charlotte, championing her side.

"Take some for both of them," said Mother. "You can go in the dogcart with Crunkleton."

As we were leaving, my eye fell on Flora Macdonald. I remembered Miss Odile's kind words. Since she so clearly admired

Flora it might be a comfort, if she *had* remembered, to see dear Flora again.

Charlotte climbed into the front seat with Crunkleton. She held the boxes and Crunkleton, stowing away the basket of flowers, saw my efforts to hoist up the fat Flora.

"Air ye a-goin' to take that slobberin' beast?" Crunkleton had no regard for Flora who chased the horses and made them nervous.

Mother must have noticed my unhappiness for she said, "Oh, take Flora, Crunkleton. She will be good, I think."

To this flight of optimism over experience Crunkleton made no reply.

Flora and I, our backs to the others rode reflectively along. Caesar was such a boring horse there was no reason to insist on sitting up front. We went through the village, no short cuts for the dog-cart.

No one could admire the village. There was the post office, catty-cornered from Smith's Hotel and across from Potts' store. Mr. Smith was the postmaster and if you thought of a vital piece of news you had omitted he would fish out your letter and what was more lend you a pen. We were not used to writing in ink and he was very kind . . . and quick with a blotter. We were violently on his side in the gossip that Charlotte and her friends dreamed up. That he was in love with his young assistant. As he was old and she was fat this seemed unlikely, but delightful.

Across the wide, dusty street was Bascom's store with its front window enhanced by a fly-specked corset covering an hour-glass figure.

We rattled across Harbison's Brook, Crunkleton sternly checking Caesar's futile suggestion of a drink at the ford.

At the Staub's Crunkleton said morosely, "Now don't stay long. I got work to do." By his own account Crunkleton was always overworked though frequently he was to be seen sitting at the stable door reading *Black's Painkiller's Almanac*. "Oh, we'll walk home," Charlotte dismissed him.

We knocked at Miss Odile's front door. I waited apprehensively. The sight of her dispelled all anxiety. More birdlike than ever she chirped and twittered, inviting us in. She looked over my shoulder and I saw her smile fade. Turning, I beheld Flora Macdonald reconoitering among the flowerbeds. Miss Odile, I thought, felt slighted at Flora's indifference.

L. P.

"Come, Flora," I called sternly. Having exhausted the charms of the flowerbeds Flora came obediently up. "Speak to Miss Odile," I commanded. It was her one trick.

Flora sat, emitted a percing bark, then offered a muddy paw. I beamed. Miss Odile gave a slight shudder, then she shook the paw and we all went inside.

Presentation of the cakes was now almost an anticlimax. But the Professor would be enchanté, Miss Odile declared. We started to his study to give them to him. Passing Miss Odile's room I saw the clock. It was one minute to eleven. "Oh," I cried, "the clock . . ."

"Go in and watch."

I went in and waited. The clock struck eleven beautiful loud bongs. The knight came out and with his jerky movement rode back and forth, back and forth. I sighed in pure enjoyment. Everything was right again.

# *Downfall*

There was each summer one entertainment upon which we could count. Hetty Cobb's birthday party.

An older sister, we knew her as Miss Cobb, would write inviting the four of us to spend the afternoon. Arguments and plans would ensue, each of us excitedly anxious to go for reasons entirely his own; none of them having anything to do with Hetty. Hetty was not exciting. She was small and drab and very shy. But this birthday party had about it the charm of discovery and change.

To us at Highlands there were two kinds of people: summer residents like ourselves and natives, the mountaineers. But the Cobbs were neither. Mother explained briefly that they were New Englanders who had come to live in the North Carolina mountains.

"Good, plain, respectable people," Aunt Florence commented.

That she, who placed so English an

insistence upon the right accents, did not ban the Cobbs', was one of those mysteries children accept. We thought the Cobb's accent queer indeed. Not like the mountaineers, yet certainly different from the Low Country tones so usual to our ears. And we knew no farm near our Savannah and Charleston homes. Plantations yes, but these were just places where fathers hunted. Crunkleton had a farm. There an apple orchard grew, dull and uninteresting. But at Hetty's, as far as the eye could see, stretched an orderly design of cultivated rows of green and gold. Best of all, at least to John and me, was the great barn. It was a heavenly place smelling of hay, horses and leather oil, and cows to be milked.

When the invitation came we could scarcely wait. Permission to go did not end our anxiety. Would Aunt Florence consider the superior age of Charlotte a stabilizing enough force? Or, horrible thought, must we have to endure the dispiriting society of Miss Jones? Without her this would be adventure.

Nothing was to be counted upon. Aunt Florence was so English. Mother was English too. But she was, one might say, less aggressively so. She was content to let us grow up in Savannah in much the same fashion as our friends. Not so Aunt Florence with Flo and John.

Word was brought by Flo that Miss Jones need not go. We sighed with relief, yet were not altogether persuaded of our safety.

At the last moment Charlotte and I encountered difficulties. It was too far, Mother said, for us to walk there and back but we held out resolutely for walking; I, protesting with heat, that I could easily walk the two miles. We won half the argument at any rate. On our return, Crunkleton was to meet us with the dog-cart where the rhododendron trail came into the road. This narrow escape we balanced against the threat of Miss Jones.

In our dogged fight for the privilege of walking we were merely following the fashion. There were few motor cars at Highlands. Unpaved mountain roads discouraged them. Walking was the thing. It was as definite a sport as tennis or golf. Indeed, the endurance to cover rapidly large areas of ground was a social asset of no little importance. Everyone boasted casually of having "done" Whiteside mountain in a morning; or mentioned in passing a walk to Lake Toxaway.

At the appointed time Charlotte and I waited outside the gate at the foot of our driveway. We perched on a rustic bench, horribly uncomfottable but fashionable then. Finally round the bend of the Walhalla road came Flo . . . but no Miss Jones. We sat back and relaxed.

John, Flo explained simply, had gone on ahead and would meet us at the trail. We made no comment. John was known to hold strong views on being seen parading through he village in the company of three girls.

Through the dusty village we went, with a pause where the mill pond crossed under the bridge. Here we would lean painfully over the sharp railing and watch the queer jerking insects that darted over the face of the water. We would spit to see the rings it made. But this time we did not tarry long. We must not be late for the party.

After a bit we came to a grassy meadow. It led us, dipping downward, to a place where grass-of-Parnassus grew, and Indian pipes. The latter we picked and held daringly in our hot, tight fists. They were, we had been told, deadly poisonous. We held them speculatively, pleasantly thrilled by this contact with danger. Then we dropped them and ran to wash our hands in the brook that flowed through the bottom of the meadow.

Now our path led up and up. My legs ached with the strain. But the penalty of admitting it was for me to be left ignominiously at home next time. I struggled along in silence.

Presently we came upon John. He was sitting on a log waiting. Evidently he felt that now without loss of prestige he might lend his presence to so feminine an expedition. At the sight of our birthday presents, wrapped in tissue paper and by now somewhat bedraggled, his face assumed a strickened look. He had forgotten his offering. Here was a situation with which we were unable to cope; there was no precedent in our small social experience. We held anxious consultation. It was too far to go back. John joined us and we hurried on. If, Flo pointed out, not averse to laboring the point, John had not been in such a hurry to get off before us, such a thing could not have happened. It served him right, she said.

At the foot of the trail, (there was a wagon road, but the trail was much shorter), we emerged on a swift-running mountain stream. This added considerable zest. There was no bridge, only a log with no handrail. John went first. He had, he said, to show us how it was done. His demonstration must have been faulty, for Charlotte, half way across, suddenly and passionately insisted she could neither continue nor return. It made her giddy.

It was an impasse. And almost within sight of the farm where Hetty and presumably ice cream and cake waited. We cajoled and threatened by turns. But to no avail. Finally when things looked blackest Charlotte solved the problem by sitting down and straddling the log. From this point of vantage she propelled herself forward in a series of leapfrog movements.

In the meantime, I had been taking advantage of the pause to examine the brook. My white shoes and socks were covered with wet sand. Even the hem of my starched party frock hung limply, discouraged from its contact with water. Charlotte, her own appearance astonishingly unruffled, regarded me with elder-sisterly disapproval. After a hasty mopping, some degree of neatness was achieved. We hung our damp handkerchiefs on a rhododendron bush to dry. All promised solemnly not to forget them upon our return.

Finally the Cobbs' house came into view. It was perched on the top of a hill and all around it were beautifully cultivated fields; and at the bottom of the rise was the barn.

When we arrived, John, empty-handed, hung back. We advanced, each holding a birthday present at arms length and presented it, as if paying admission. We saw there were other children present, neighbors of the Cobbs' whom we never identified.

The Cobbs' front room was fascinating to us. There in the mountains of North Carolina was as perfect an example of a New England parlor as one could find Down East. The furniture was slick black horsehair that defied bodily impression. There were embroidered texts on the walls. On the mantelpiece sat rows of sea shells. These the elder Cobbs would hold to Hetty's ear, pathetically anxious that this child of the Southern mountains might hear her rightful heritage . . . the boom of surf upon New Enhland's shore.

Hetty's elder sisters were much in evidence. One felt they held this to be an occasion. The guest of honor sat meekly in the background in such a visible state of shyness that we averted our eyes. Her sisters would try to make this her party. But they soon gave up. Hetty was born to be overlooked.

Presently the gramophone would be turned on as we listened rapturously. It was true we had one at home; but not to compare with this. Ours was an uninteresting cabinet that under the guidance of some grownup played mostly things called overtures. This was a box with a large horn through which came the most astonishing and gratifying entertainment. The Cobbs' choice of records we considered infinitely superior to that of our parents.

"Cohen on the Telephone," "Two men to mend a damaged shutter. No not shut-up Shutter," to us was the ultimate in humor. But reserved for the last, and indeed a fitting climax, was a horribly realistic rendition of "The Battle of Santiago."

We sat tense with excitement as this dramatic incident unfolded. The record started with Morning Prayers on deck. Devotions were quickly terminated when a ringing voice announced, "The Spaniards are coming up the harbor, sir." We held our breath and our eyes were glazed in anguish while cannons roared and shots sounded. Through it all could be heard the moans of the wounded . . . singularly loud and healthy in effect . . . but eliciting from its hearers the tribute of goose-flesh. Finally the shots grew fainter as a noble voice admonished the victors. "Don't cheer, boys. The poor devils are dying." The reaction was such we sat motionless, without a word. The only sound was the scratching of the needle on the finished record. No third act could approach it in sheer dramatic quality.

After this excitement we were calmed down with stereoptician views. Each in turn set the hard bridge of the instrument upon his nose. Owing to the fact that the Miss Cobb who adjusted the vision was farsighted we had some difficulty in distinguishing Niagara Falls in the Winter from The Grand Canyon.

But to the Cobbs the chief business of the afternoon was food. Real New England cakes that glistened with icing; served in wedges the like of which we had never seen. And there were doughnuts and maple sugar candy and lemonade made with cold, spring water. We fell to with gusto; but Hetty minced at her plate.

Later, surfeited with food, we sat upon the porch enjoying the short period of calm the elder sister permitted for digesting. It would not be for long. Continuous action seemed in her eyes the tuff of which parties were made.

As we played we could see the barn. A horse was tethered outside. John and I edged further and further away from the party. The exigencies of the game covered our escape and soon the brightly determined voice of Miss Cobb, counting in fives for hide-and-seek, was dimly heard and quite unheeded.

Mr. Cobb was chopping wood; fire wood for the kitchen stove. We had wood stoves in our own kitchens, but their fuel usually came in a wagon, and Mother bought a cord at a time. We watched fascinated. He'd put a small piece of log on the chopping block standing straight up and hit a wedge into the center of it with the back end of his axe. It would fall in halves. These he chopped into neat pieces a foot long and two inches wide, slick and smooth. "It's good oak," he said when he noticed us.

We went into the barn. It was a cow barn and smelled like cows, not horses and was muckier underfoot. Mr. Cobb went on chopping wood.

When he paused to pick up the falling pieces and stack them in a tower like a little log cabin, John asked him the name of the horse, which proved to be Ralph.

John said Ralph was a funny name for a horse, "Sounds more like a boy."

"So?" murmured Mr. Cobb.

"Now ours," importantly, "are named First Over and Little Starlight."

"So?" observed Mr. Cobb.

John regarded the horse critically. "Bit spavined," he commented at length.

Mr. Cobb appeared not too disturbed. "So?" He went on stacking. But I, who loved anything equine, flung my arms about Ralph's neck in a passion of apology. Ralph endured the embrace with a rolled, resigned eye. Frankly, his look seemed to imply, there is no need for such demonstration.

"May we ride him?" I asked.

"Oh, may we?" begged John, suddenly humble.

"Well, I don't see why not." Mr. Cobb

brought out a sheep skin and girthed it round the horse's back.

"Can you ride on that?"

John had regained his usual aplomb. "Oh rather. We learned on a sheep skin. For balance, you know."

"So?" Mr. Cobb lifted his brows and peered over steel-rimmed spectacles. John appeared slightly abashed and I reflected that though Mr. Cobb did not use many words he conveyed what he meant when he spoke.

"I'll ride first." John was never abashed for long.

He was lifted up and rode away exclaiming the while that one could not properly post without stirrups.

"Talkative fellow," Mr. Cobb jerked a thumb after the departing rider.

"He's been to boarding school . . . in England," I replied with awe.

"So?" Mr. Cobb was not impressed. Then he looked at me and dropped one eyelid in a slow wink. I decided that Mr. Cobb was a very nice man indeed.

John seemed to be taking a long turn. I peered down the road anxiously. Finally I saw him coming.

"Get off, get off," I screamed. "It's my

turn now."

"You couldn't manage him."

But Mr. Cobb seized him and swung him down. "I was just going to get off," he complained in an injured voice.

Once up I felt Ralph was a very tall horse. We started slowly, but under the critical eye of John I felt compelled to a brisker pace. It was difficult without stirrups. I went down the road by the kitchen garden and paused at the spring house. Here, as I bumped along my hair ribbon began to slip off. I reached for it, too quickly, lost my balance and over I went to land, in the soft, tall grass by the roadside. I lay there completely astonished at my sudden change of position. Ralph sniffed curiously at me, blowing a dusty sigh in my hair and tickling my cheek with his rubbery lips. Fortunately my ignominy was hidden behind the spring house. I scrambled up and brushed my dress. A large grass stain adorned the skirt.

The distance to Ralph's back seemed enormous. Yet to go back on foot was unthinkable. I stuffed my hair ribbon down my front and considered. By the spring house a lichen-covered rock jutted from the ground. I led the patient Ralph close and climbed to a height where I could mount. He seemed surprised. I was tugging very hard at his mane.

But we went back in style, my face very red and my knees, beneath my grass-stained skirt, keeping him at a steady, impressive canter.

From the house there came a loud "Who Ooo" through cupped hands. Our absence had been discovered.

John prepared to take his turn, but Mr. Cobb, somewhat to our surprise, appeared to take the calls seriously though all he said was, "Seems to want you."

"Oh *that*," John airily attempted to mount; but he was dissuaded and Ralph removed.

The cries became more insistent now with pessimistic conjectures as to reckonings at home.

"Well, p'raps we'd better go. Old Ralph isn't bad, Mr. Cobb, for a horse that's not been schooled."

"Ralph," I exclaimed with indignation, "is one of the nicest horses I ever knew."

At the house preparations were made for departure. A round of handshakes were given . . . there was a very limp one from Hetty . . . and the formula "We had a nice time" repeated.

The woods were shadowy with twilight and the evening sounds of the birds like hushed breathings. I clung close to Charlotte's protecting side. We negotiated the log easily, Charlotte evidently forgetting her fright in the larger menace of approaching night.

Suddenly I remembered our handkerchiefs, still tied to the rhododendron bush, fluttering in the growing dark, alone. I pictured them waiting there for us who never came. They would be there through the late summer and in the winter covered with the snow I had never seen and could only faintly imagine.

"We'll get them next year," comforted Charlotte.

But John scoffed. "Wild beasts will eat them."

I burst into tears. Home seemed so far away and wild beasts so imminent.

Then the trail ended and suddenly we were in the high road. And, oh joy unspeakable, Crunkleton had come to meet us. He sat there in the high dog-cart and grumbled at our lateness. But we burst upon him with shouts of delight.

"Thought you-uns was the ones as wanted to walk there *and* back."

But we ignored such want of tact.

John and I got up in front with him. Charlotte and Flo took the rear seat where they sat with their backs decorously to us.

"I'll drive," offered John. "After all, Crunkleton, you're probably tired with all this waiting . . ."

But Crunkleton merely spoke to Caesar and we drove off in a forceful silence.

After a bit the fat cob slowed down. "I'll just touch him up for you." John reached for the whip and in some way got his finger pinched in he whip socket. I eyed him with satisfaction; but Crunkleton shook with silent mirth. Then, "Lay-overs to ketch meddlers," he observed. It was one of his favorite remarks but we never knee what he meant by it.

After that we drove in drowsy silence. Nothing broke the evening stillness but the occasional click of the horse's hoof striking a loose stone. Around us the woods were dark and strange. Far in the sky the new moon hung; even the brash, informative voice of John was still. I sat there, legs sticking out over the leather seat, eyes wide with the glory of the party, in blissful

L.P.

contentment . . . never for a moment believing that life could be anything but the same, tomorrow and tomorrow and tomorrow.

As we rode along lamps presently began to show in the darkness. We passed the village and up to our mountain road. Crunkleton stopped at our gate.

"Yer ma mought be a-gitin' anxious. You two young-uns cut along to the house." He would drive the others home.

Obediently we climbed down and stood for a moment in brief judgment of the afternoon. It had been fun at Hetty's.

We would go back . . . . soon.

But we never did. Not again that summer.

# *Haskell and the Bishop*

I found Haskell in his pantry making mayonnaise. Casting a glance at the kitchen door I said, "Sarah know you're doing that?" Haskell's insistence upon this task was the spark that ignited the feud smoldering between them. Mother, an anxious mediator, was spurred by the knowledge that, either one a treasure, together they were priceless. I, on the other hand, found their clash of arms exciting.

It all stemmed, Father morosely declared, from Uncle Peyre's travels. "But it gave him an air," Mother said.

"Which one?"

"I meant your Uncle Peyre; but there is no denying Haskell has a manner."

Uncle Peyre, taking Haskell, had spent a year in England and France. Being English, Mother recognized certain echoes of this sojourn. One of them was Haskell's use in the large sense of the term "my pantry." By this he meant not so much the physical limits of his domain as the duties and prerogatives it conferred upon him. The making of mayonnaise, unfortunately, was one. It was also a skill cherished by Sarah. At home, also on the Continent, Uncle Peyre had Haskell instructed in the art of making salad dressings in the French manner. It was an accomplishment that brought him acclaim in Charleston; he had no idea of relinquishing it now to Sarah. "Mayonnaise go wid my pantry;" and there the matter rested for him. It was, however, a challenge Sarah continued to meet.

At my question Haskell turned his head a bit apprehensively toward the kitchen door, then, "Sarah ain' know," he said and laughed "Ki, ki."

"Haskell, why do you always say 'ki, ki'?"

Haskell looked displeased. I realized he felt my manner at fault and said quickly, "Haskell, do you know what a bishop is?"

If I thought thus to placate him, the more fool I. "Does I know what a bishop is... Does I know..." He stopped, speechless with outrage. Finally he said with dignity, "Ain' you know a bishop one of me an' yo' Uncle Peyre's bes' fren's?"

I was surprised. I had heard a lot and been fascinated by what I heard of

Haskell's and Uncle Peyre's lives in Charleston. I had not, however, got the impression that Uncle Peyre's associates were of the higher clergy.

"I didn't know Uncle Peyre knew a bishop."

"All yo' Uncle Peyre's fren's high class."

"Was this bishop a Charleston one?"

"He live up Nort' but he use to come to Charleston winters."

"He must have been retired."

A broad grin spread over Haskell's face. "He had a fren', ole Judge Lucius Wade. We use to have 'em to dinner once or twice a year. An' I really mean it *was* a dinner."

"Did they like to eat?" I did not connect bishops with the pleasure of the table.

"Judge Wade sho did. It did my heart good to see that man lay to. 'Ain' nobody make turtle soup and shad roe lak you,' he'd say, 'yo' whole dinner a poem.'"

I digested this in wonder. I had a passion for poems and for learning long, story-telling poems and was at the moment busily engaged with *The Blind Girl of Castel-Cuille*. I could not quite see how poor Margaret's troubles could be compared with a dinner. But I knew better than to question Haskell's reminiscences. "Did the

Bishop like to eat too?"

Haskell laughed. "He ain' so fond of food. But that man really liked his drink."

"Drink!" I cried in amazement. "You mean the Bishop drank a lot?"

Haskell looked at me in pity. "Folks don' have to get drunk to like good liquor. The Bishop ain' that kin' of a man. He what Mr. Peyre call 'a conisoor.' W'en he coming Mist' Peyre tell me, 'git out us oldes' cognac. The Bishop comin' to dine.'"

I remembered my news. "Father has a friend who is a bishop. Do you suppose it's the same one?"

"Dis man daid."

"Oh. Well, this bishop is Miss Hattie Gaillard's nephew and he's coming to stay with her. Father wants him to stay here so he can go fishing. But he says," I quoted, "'It can't be arranged.'"

"Won't get no fishing do he stay nex do. He'll be too busy handing her medicine an' puttin pillows behind her shoulder, an' tending to what that noise is, seeing does her horse stay in a walk, if they go for a buggy ride. That lady sho scare easy."

"I heard her say she will not ride in an automobile, not in these mountains. You know, Haskell, there are *five* automobiles up here this summer. I wish we had one."

"You rather have automobile than *horses*?"

"Well, no, but it would be nice to have both."

"Moonlight and Magnolia git right up on their hin' laigs. This Bishop fren' of yo' father, he a Charleston man?"

"No. He and Father were friends at college; he lives in the Middle West but Father says he used to stay with him in Charleston . . . at Christmas holidays. He came to Savannah last year and preached but we hardly saw him."

Haskell had begun to drop olive oil into his bowl. Now he stopped and put down the bottle. "Yo' mean Mist' Jimmie Hamilton got to be a *Bishop*?"

"Yes. Do you know him Haskell?"

"Ki, ki. I certainly does. He came roun' to the house two, three times with your father." He aughed. "I gave him a tie once."

"For Christmas? Was it a Christmas present?"

"No suh. This a necessity. He an' yo' father came roun' to go to some affair with Mist' Peyre an' I let 'em in. Mist' Jimmie

Hamilton, he sort of a reckless dresser an' I see his tie all rumpled . . . and crooked too. It wasn't fitten he go lak that . . . not with Mist' Peyre. I ask him, can I fix it for him." Haskell gave an indulgent chuckle. "T'wan't no fixing that tie. So I got him one of Mist' Peyre's own an' put it on him. Never did get it ack, neither. Well, well. He sho wan't no bishop then."

"Father says he didn't go into the ministry until he'd been out of college a few years. But he's real young to be a bishop all the same. He's coming in the morning in is car. I can't wait to see it."

"Better not take no automobile to Miss Gaillard's. It'd scare her *too* much . . ."

Next day after lunch I was on the verandah absorbed in learning the whole of *The Blind Girl of Castel-Cuille* to confound Charlotte upon her return from Blowing Rock where she was visiting a friend. Suddenly she had passed another one of the invisible but real milestones of growing up and become knowledgeable as befitted one of superior years. Left behind, I was puzzled, but admiring.

Mother and Father came in from golf and joined me. I turned a shining face. "I've learned a new poem. It's out of a book

Charlotte got for her birthday. Longfellow."

Mother said, "I don't care much for Longfellow."

I waved her aside. "Oh, it's beautiful . . . and so sad. Just listen:

"Village girls in robes of snow
Follow, weeping as they go;
Nowhere was a smile that day,
No, ah no: for each . . ."

I broke off. "I can hardly say it without crying," I said ecstatically.

Father sighed. "Later."

The gate clicked. We looked up to see a man striding up the path from the direction of Miss Hattie Gaillard's place next door. Father got up in pleasure. "Jimmie! By all that's holy." Then he laughed. "Appropriate, eh? But you haven't got on your dog collar."

It was the Bishop. I noticed he wore an ordinary tweed suit and collar and that his tie was slightly to one side. I thought of Haskell. Apparently the Bishop was still a reckless dresser.

Father and the Bishop rushed at each other and shook hands at length in a vehemence of greeting. They were laughing and shouting and seemed very glad to meet again. I watched them with interest but little sympathy. Why will men act like boys? I thought critically. I didn't know about the Bishop, but Father was old. He'd have to be to be our father.

The Bishop greeted Mother warmly, if with less enthusiasm; then he turned to me. "How you have grown." I gazed at him coldly; The words; "it was to be expected," hung in the air between us but were never spoken. Then the Bishop gave me an apologetic smile and went to sit beside Mother.

"It's very disappointing, Emily not coming," she said, "but I hope you're going to be able to spare us a day or two. George says the fishing is good and he's saved secret places for you."

The Bishop groaned. "Nothing I'd like more, but I'm afraid I'll have to stay with Aunt Hattie. We find we have to leave Flat Rock for home sooner than we had planned. Emily's cold was better, but a cousin from Charleston arrived . . ."

Father laughed. "I remember your saying, 'Charleston is so full of relations!'"

"It still is; just when Emily was well of

her cold . . ."

"Haskell says," I interrupted, "that when he has a cold every tooth in his haid aches." When I quoted Haskell I was apt to do so literally.

"Is that so?" asked the Bishop. "Is that a Mr. Haskell from Charleston?"

"He is from Charleston and he holds it over Sarah. (She's from Savannah) But we don't call him *Mister*. Hudson does . . . he's the mountaineee who brings our eggs . . . and I think it's only fair as Haskell calls *Hudson* mister. But . . ."

"Edith," said Father, "Please ask Haskell to bring us some sherry and biscuits. Tell him the bottle I got out yesterday."

The Bishop looked enlightened. "Charleston's full of Haskells too. Is this the one who used to look after your Uncle Peyre? I remember him well."

"Haskell," I cried when I found him, "the Bishop's here. And he remembers you!"

"Of course."

"But he hasn't got on a Bishop's collar. It's an ordinary one and his tie is still crooked."

"Is that a fac?" And Haskell laughed

heartily.

Shaking hands with Haskell, the Bishop asked him if he remembered how he'd fixed him up. "I don't believe I ever returned that tie."

"It was a pleasure, suh, a pleasure." Haskell was beaming. "We couldn't have you going to the ball that way."

"You know I think I owe you more than a tie. It was at that party I met my wife. And I don't believe Mrs. Hamilton would have thought much of me before you set me straight."

"She didn't come with you, suh?"

"No, she had company and I can only stay a day or two. I'm next door at my wife's aunt, Miss Gaillard."

Pouring the sherry Haskell looked sad. "Fish biting good," he said.

"Couldn't you at least go fishing tomorrow?" Father urged. "I do know a special place . . .' He stopped and looked out across the lawns and fields. "It will be good weather for fishing." It was a radiant day with all the colors of noon in clear relief. Far in the distance the Nantahala mountains were misty and blue; nearer at hand the Fish-hawks thrust three, bold heads against a sky washed with light.

Father turned and looked toward Satulah towering solid and close at hand. Clouds were crowding its brow, delicate now, but suggesting in their restlessness heaviness to come. "Yes," he repeated, "good weather for fishing. Those clouds over Satulah mean it will be overcast tomorrow. You'd better come, Jimmie."

"Afraid not. The old lady has things pretty well ordered. This afternoon I am goingup to call on old Dr. Elliott. His wife was another of Emily's aunts and tomorrow is mapped out just about the same."

"Did Miss Hattie like your car?" I was curious to see it.

The Bishop laughed. "Not too much. I gather she does not care for them. The surprising thing," he turned to Father, "is that she has agreed to drive down to Cashier's Valley tomorrow . . . to see another one of Emily's relations. Nothing but the thought of seeig a relation would entice her, I'm sure."

I happened to be looking at Haskell just then and I saw him purse up his lips the way he did when he was thinking. He recorked the sherry thoughtfully.

The Bishop would not stay for lunch and got up regretfully to go. "I'll see you again.

But really I can't desert Aunt Hattie to go fishing. If Emily could have come – "

That afternoon I was on the verandah still learning my poem. It was deliciously sad. I found enjoyment in the fact ... and in the glorious afternoon. My gaze traveled out across our lawns, over the road and into Mrs. Hill's meadow. At he bottom of the meadow were tall hemlocks; far away over their heads I could see the Fish-hawn mountains. Mrs. Hill, another Charlestonian, kept her lawns beautifully shaven; but she left her meadow tall with daisies, blackeyed Susans and Queen Anne's lace. I never looked at it, brimming with summer, without saying to myself ... or to anyone who would listen:

When daisies pied and violets blue,
And lady-smocks all siver-white,
And cuckoo-buds of yellow hue
Do paint the meadows with delight...

"Paint the meadows with delight." That was it exactly. I was continually being struck by the satisfactory way poets echoed my thoughts.

Mother had taught me that poem. She would say lines and verses she liked. Once, walking to the dentist at which I was the most arrant coward, she taught me a useful bit out of Henry V:

"Hold hard the breath, and bend up every spirit
To his full height. On, on, ye noblest English
. . . Show us now the mettle of your pasture."

It bolstered me. After all, as Mother pointed out, I was half English. I often found comfort in it as, for example, when confronted by especially awful irregular French verbs.

I admired the poems Mother showed me; but thoughtthey lacked excitement. More thrilling were the narrative poems I found in *The Children's Hour*. I knew and gloried in *The Relief of Lucknow*, *The Pied Piper of Hamelin* and several ballads of derring-do. I had just discovered Longfellow.

Haskell came out on the verandah and I cried, "Haskell, listen to this . . . it's wonderful."

"And by suffering worn and weary,
But beautiful as some fair angel yet,
Thus lamented Margaret,
In her cottage lone and dreary . . ."

"Nem min' her, ain' that the Bishop going by?'

"Yes, it is. Do look, Haskell, he's got on plus fours, to climb the mountain I suppose. He doesn't seem much like a bishop to me." Then suddenly I thought of something. "Haskell, I believe you've been standing in the living room watching for him to go by."

He ignored this. "W'ere yo' mother and father? Gone out?"

I nodded.

Haskell shook his head. "Too bad."

"Why is it too bad, Haskell? What do you want with them?"

"I thought..." he sighed and topped. Haskell knew the value of suspense.

"Oh Haskell!"

"Well, do seem to me the Bishop yo' father frien' an' all ... somebody ought to do something polite for him."

"How?"

Haskell appeared to ponder. "Well, you could take him some flowers."

"*Flowers*? You don't take flowers to men, Haskell."

Haskell looked superior. "Yo' ever take any flowers to ladies?"

"Well, no . . . but . . ."

"All right," Haskell said triumphantly, "shows you don' know nothin' 'bout it. Let's pick a nice bunch of flowers an' take 'em to Miss Gaillard's for the Bishop."

"Right *now* take them? If they're for the Bishop why not wait till he comes back?"

"Right now!" he went into the house and came back with Mother's secateurs, "Come on." And I followed him out into he garden.

"How 'bout some nice roses?" slashing about with the secateurs.

"Do you think Mother will mind?" I knew she was particular about her roses.

"She'll be pleased we thought of it. She'll say, 'Now w'y ain' I thought of that?'"

"Well," I took the roses Haskell had cut, "Pick them with longer stems." If we were going to take Mother's roses we might as well do it properly.

"That's got it," he said presently. "That's a bewdyful bookay," thrusting them at me.

"They've got awful thorns." I took them gingerly. "Tell you w'at. I got to see Maner anyway." Maner was Miss Hattie's servant. "I'll go 'long with you and carry 'em for you."

We went out the side path, through our little gate and across the road that separated the two places. The air was warm and still and somnolent with the hum of bees. Miss Hattie's path wound up through the rhododendron thick with bloom, their luminous color made brightness in the shaded walk. I reached out and picked a flower from a cluster. It clung stickily to my fingers.

Emerging by the house we saw Miss Hattie sitting on the verandah knitting. She nodded brightly to us and Haskell shoved the roses toward me "Go give 'em to her," he ordered.

I walked up the steps feeling pretty silly. Haskell was a stickler for manners: I had the greatest faith in his *savoir faire*, but taking flowers to the Bishop seemed odd. Haskell had stopped to greet Miss Hattie. Now he went kitchenward and I was left to cope with the situation.

"Mother sent these," I said ambiguously. I did not mention the Bishop. Miss Hattie accepted my offering with pleasure. "How kind of her. Maner is not good with flowers . . . he really doesn't care for them. But he has a nice vegetable gyarden. I must give you some lettuce."

I started to leave. "Can't you stay?" I shook my head and she said hospitably, "Well, come along then. We'll go to the gyarden and see if Maner has some lettuce for you . . . he might have tomatoes too."

We walked down the steps and round by the back of the house. We found Haskell and Maner deep in talk. "Maner," said Miss Hattie, "do pick some nice lettuce for Mrs. Inglesby . . . and tomatoes if they're ripe."

Haskell turned and came up to us. "It's a great pleasure for us, Mam, to see Mist' Hamilton again."

"Yes," Miss Hattie said with satisfaction, "he's a bishop now, you know."

"I say all along that gemman goin' to *be* somebody."

I suspected Haskell had never given a thought to be Bishop since Charleston days but mentally I applauded his social sense.

"He say," went on Haskell admiringly, "you is going to drive with him to Chashier's Valley in that fas' autymobile he got. My, you is brave, Miss Gaillard. I do declare you is, rally. That road would make a rale autymobile-lover scary, an' I didn't think you cared 'bout 'e so *much*."

Miss Hattie looked alarmed. "Is it a bad road?"

"Bad! Great Gawd! That road make you take yo' life in yo' han'. It wind and wind and all they is on one side," he said with a grisly laugh, "is air . . . jus' air."

Miss Hattie shuddered. "I don't much care for cyars. Maybe I ought not to go . . . if the road is as bad as you say." She looked so pitiful I felt sorry for her.

"No'm. You go on, Miss Gaillard. T'ain' many ladies would do it. It'd tarryfy them. Then when you git back . . . *do* you git back . . . you can let 'em all see how venturesome you is."

"Venturesome and foolhardy," Miss Hattie moaned. "I never did care for cyars up here. They're bad enough in Charleston."

"Yes'm, Charleston is different. It's safe!"

"I just can't go," sad Miss Hattie with determination. "I am not going to ride in a cyar on that road."

"I don't blame you, Miss Gaillard. Was I you I'd jus' stay home."

Miss Hattie hesitated. "But I said I'd go."

Hasell looked sympatetic. "Tell you w'at

you do, Miss Gaillard, you jus' stay in the baid with a haidache. That'll fix it for you. A haidache sho would be safer."

"I don't feel at all well, not at all. Perhaps I oughtn't to go. But I told my nephew . . ."

"Let them w'at likes risk they necks in autymobiles." He paused to observe Miss Hattie's flinch, then added, "Ladies is more delicate."

Maner came up to say there were no tomatoes ripe enough for icking. He handed a basket of lettuce to Miss Hattie who eyed it in an abstracted manner as if wondering why it wa there. Apparently recollecting, she gave it to me declaring Maner would gather tomatoes for Mother when they were ripe.

I thanked her and said goodbye. As I started off I caught a sternly admonishing look from Haskell. Going back I offered my hand to Miss Hattie and dipping my left knee slightly gave the ground behind me a vigorous kick with my right fot. This version of a curtsy having satisfied both Haskell's and my ideas of proper deportment we took our departure.

I walked along in silence. Where was Haskell's tact? Where the polish that

Charlotte and I so admired? Charlotte said she thought Haskell was as much a man of the world as any friend of Father's. Not at all sure what she meant, I gathered by her tone it was something impressive. Wait 'til Charlotte learns how gauche he had been. Never one to keep my thought to myself for long I blurted forth, "Haskell, you *frightened* Miss Hattie."

"Oh, oh," Haskell gave an apologetic, clucking sound.

"You scared her so much about that road that I bet she won't even go."

"You think she won't?" Haskell knotted his brows in thought.

"I certainly do. And what will the Bishop do then?"

"Auty chos," said Haskell.

"What?"

"That's something yo' Uncle Peyre use to say at times. It's some kin' of foreign talk. Aunty choses," he repeated.

"Oh! Autres choses . . . other things. But what other thngs can the Bishop do, Haskell?"

"Well," Haskell said blandly, "wait and see."

At breakfast next morning Father looked out the window with satisfaction. "No haze this morning. It's going to be overcast. Good weather for fishing."

Suddenly someone called, "Good morning." It was the Bishop. He has a good voice for chanting, I reflected. It was a clear resonant baritone that did much, I felt, to offset his clothes, but he looked less than ever a bishop. Father eyed him hopefully. "Surely those garments were not donned for a visit to a Charleston relation?"

"Hardly," said the Bishop, strutting. "I was hopping to catch you before you got off this morning. How do you like my waders? Got 'em in England."

"Just like mine, but I got mine at Abercrombie and Fitch."

"Won't you have breakfast with us?" Mother asked.

"Thank you, I will. I passed Maner taking up a tray of weak tea and toast to Aunt Hattie. However, I suppose he would have done better by me."

"Then you aren't going to Cashier's Valley?" Mother asked.

"No," said the Bishop cheerfully. "Aunt Hattie has a headache . . . and it seems she was advised against the road."

Haskell, bringing a chair up to the table, wore his usual impassive expression. He doesn't look at all sorry, I thought, for scarng Miss Hattie like that. And although the Bishop's face was serious, his eyes were twinkling. After breakfast the Bishop went home to get his car. I was going down to the village to mail a letter to Charlotte but I waited on the verandah for his return. His automobile was full of shiny brass and it made a tremendous oise. I thought it enchanting. The Bishop saw my admiration. "Would you like a short ride?"

"Oh, could I go down to the village? I have a letter to mail."

Father came out of the house with his rod and creel. His hat was full of brightly colored dry flies and he was opening his book of them affectionately. "I think the Silver Doctor this morning," he said eyeing the weather. He looked at me in surprise. Obviously I was waiting to get in the car. "You can't go, Edith. Not this time."

The Bishop said I was only going to the village. It was a two-seater car with scooped out seats with curved sides. The Bishop took the wheel. Father sat in the other seat and held me in his lap. We went down the driver rather fast and I did wish Charlotte could see me now. Out of the gate we sped along the road to the village in an explosion of sound and some smoke. Dogs barked, cows got angrily out of the way, their bells jangling in rage.

I was exhilarated and as usual at such times full of talk. The wind, whistling across my face, reminded me of the book I was reading ... my latest "favorite." Above the roar of the engine and the wind I laned forward and shouted to the Bishop. "Have you read *The Wind in the Willows*?"

"Don't think I have."

"It's wonderful. Mr. Toad had an automobile, too. He liked it better."

"Better than what?" shouted the Bishop.

"Boats," I roared.

"I say," Father interposed, "can't you postpone this literary discussion until some other time? I don't hink my eardrums will stand it."

Having been put down in front of the Post Office I looked about hoping some contempoary had seen my glory. But there was no one of interest. Glancing down at Charlotte's letter I thought what a lot she would have to hear when she came home. I had told her about my new poem; I did not, however, feel that it was news that would hasten her return. She was a patient rather than an eager listener.

She would certainly wish she had been there to see the Bishop's car; and she would be dumbfounded to hear abut Haskell's extraordinary lapse of finesse. By his tactlessness he had kept the Bishop from going to Cashier's Valley in the car with Miss Hattie. He had been all set to go.

But then the Bishop, I decided after further thought, really would rather go fishing with Father so . . .

Suddenly my innocence was pierced by a new and startlin idea. *Was* Haskell being tactless . . . or was he. . . ? I stopped short in amazement. Haskell *meant* to do that! Haskell could do anything he wanted to. Haskell, I recognized with awe, was a Man of the World.

# *A Thoughtful Child*

Every summer has its fashion. There were, for John and me, always horses; but each year brought its lesser interest. Once it was damming brooks. Each day we would work like beavers bringing mud and rocks to stem some crystal mountain stream. Our legs nearly numb in the icy water, we would work for hours only to find the dam we'd left holding water at nightfall had been undermined by crawfish by morning. Other years we caught and mounted butterflies; dug beryl and pried garnets from the rocks on the mountain sides. This year was the year of the wigwam.

Father was teaching us to build wigwams. He had made them, he told us, as a boy on the plantation; but we thought ours would be nicer. At Charleston there would not be the sweet-smelling ferns that grew so plentifully at Highlands.

John was away visiting a schoolmate at Flat Rock. I rode every day; but with him away I had more time for other activities such as wigwams.

Every day Charlotte and I . . . and Flo when available . . . followed Father through the spinney at the head of our property. We cut the saplings Father thought would be right. My hands were better on the reins than holding a hatchet; I hacked away, never hitting the same place twice. Father said it might be better if I just helped carry the poles.

We chose a place far away from the house to put the wigwam, down in a corner of the fence at the bottom of the lawns. The independence suggested by this isolation pleased us. Father drove the poles into the earth in a fairly large circle; then he tied the ends together at the top with a rope. With twine he made a sort of lattice round the poles. Through this web of twine and poles we wove the ferns. We had gathered them in the woods, piling our wooden cart high with the fragrant loads. There were enormous cinnamon fern, royal fern, hay-scented fern and bracken. Charlotte and Flo, with some importance, identified them. It was enough for me that they should smell so sweet and be so concealing; we were in another world. It was lovely there, green, scented and remote.

I wanted to bring my Magnolia to see it. I would just lead her half way in . . . so she could see. But Charlotte objected. Maggie would eat the fern. Wasn't Flora Macdonald enough? Flora thought so. Father said she was an extrovert. Fat and untidy Flora had a touching belief in the friendship of the world . . . a belief that perhaps a fortunate insensitivity sustained.

One morning after breakfast we were on the verandah. Mother and Father were about to leave for golf and Charlotte and I had, of course, to go to the wigwam. Crunkleton came up from the stable with the mail he had brought from the village.

Reading a letter, Mother looked up. "George, Ella and Alice are coming, they'll be here on the fifteenth. That's next week."

"Good," said Father. "I don't suppose poor Ella gets out of Charleston much now."

Mother turned to Charlotte and me and said, "Your Father's cousin and her little girl, Alice, are coming for a visit. It will be nice to have another little girl to play with, won't it? Alice is just about Edith's age."

"What will she ride?" I asked.

Mother said a trifle severely, "You'll let

her ride Magnolia, won't you? It would be selfish not to."

"Well," I said not very enthusiastically, "she can . . . some. But I'm getting Maggie ready for the show. Charlotte doesn't ride much. Why can't Alice ride Moonlight?"

"Moonlight's pretty skittish with strangers," said Charlotte, "but maybe Alice is good with horses."

"I don't think she will be," Father put in. "She wouldn't have much chance . . . except on the plantation. Eugene used to take them there; but there wasn't anything to ride but marsh tackies."

"Who is Eugene?"

"Eugene Dawson. He was Alice's father," Mother said. "But he died last year. Alice must be a nice child." She picked up the letter and read: "Alice is the greatest comfort to me. She is such a thoughtful child."

A contemplative silence greeted this, then Father looked at his watch and said he must not keep his match waiting. They went off with their golf bags and I went in search of Haskell.

Haskell was in his pantry drying dishes. From the kitchen came the high wail of Sarah's singing:

"Sometimes I feel lak a motherless child."

It was known that when she sang spirituals Sarah's feelings were hurt. Coming, as he felt, from the *haut monde* of Charleston, Haskell was inclined to a feeling of prestige. Sarah was not one to let Savannah be slighted. Between them there was a wary and hostile truce, broken sometimes by acrimonious rebuttal. The atmosphere now was charged. Ordinarily I found these clashes of arms exciting, lending to each combatant a sympathetic ear. Today, however, I had things on my mind.

Haskell looked toward the kitchen then lifted his eyes heavenward. But I evaded this opening.

"Haskell, I have a cousin coming to visit me. Her mother says she's a thoughtful child."

"Is that a fac'?" Haskell finished rubbing a spoon, then he observed. "Yo' might study on her manners."

"Yes. But suppose I don't like her?"

"How 'bout is she gonna lak *you*?" Haskell said and laughed. He had a robust sense of humor. "She a very lady-lak little girl, what she gonna think of your britches and all this horse?"

"I'm going to let her ride Maggie ... sometimes. But Father doesn't think she likes riding. She's from Charleston."

"Charleston? Who her pa?"

"He was Mr. Eugene Dawson. But he's dead."

Haskell pursed his lips. "I knew him. He was one of the tea drinkers."

"Tea drinkers?"

"You remember I tol' you. Yo' Uncle Peyre used to give a big tea party once in a while. Lots of ole ladies would come. Some of the gemmans would take a cup of tea. This Mr. Dawson he a tea drinker. Never came in the libr'y for a bit of cheer, like the other gemmans."

He gave a sigh. "Now he daid. Po' man." Evidently he thought Cousin Eugene had missed out all round.

In the kitchen Sarah was still singing.

"... a motherless chile.

A long way-a-y fum home."

I went to join Crunkleton at the stable. "I want to ride," I said, sensing a troubling of our calm days.

Next week, on the day Alice was to arrive, Charlotte and I dressed for the afternoon with care. Alice's arrival was an event of anxious anticipation. We were very excited.

The nearest train to Highlands was forty miles away and the final lap of any journey was made in Dillon's Ford car. It was a conveyance that heralded its approach. At about five o'clock we could hear it chugging up our driveway. Mother and Father went out on the verandah. Charlotte and I hung back a trifle diffidently. Not so Flora Macdonald. She flew out with piercing barks, then subsided on the top step in a welcoming position. Father helped Cousin Ella and Alice out and they came up the steps. Immediately Flora bustled up grinning her fatuous grin. Alice drew near to her mother. We could not have been more astonished. We were aware that grownups sometimes found Flora expansiveness trying; but Alice was the first child to view our dear familiar askance.

Rather dashed, we came forward at Mother's look and were introduced. Alice's hand lay limpy in our energetic handshakes. She was a pretty girl with light curls and blue eyes. We stood awkwardly, then in a burst of hospitality I suggested we go see the horses. I could imagine no greater treat. As no enthusiasm was evinced by Alice, Charlotte then inquired if she'd like to see the wigwam. But Mother called us in to tea. Tea, she apparently thought, was indicated.

Alice was no hilarious guest; but we found her pleasant. Flo came over every day and we four spent most of our time in the wigwam. We were allowed to build a fire near the entrance where we cooked lunches in a heavenly feeling of independence. Haskell helped us to obtain supplies and we ate half raw potatoes and burnt ham with enthusiasm and unbelievable digestion.

The days of Alice's visit went agreeably enough. She did not want to ride. When Charlotte and I rode .. it was unthinkable to me that the sun should descend upon a day without riding . . . Alice would employ herself round the house. To Haskell's wrath she wanted to help him with his tasks. It seemed a poor return for her thoughtfulness that Alice's efforts in his behalf should be unappreciated. But we were forced to conclude they were.

One day we were on the verandah waiting for lunch. Father had just come in from golf and was having his whiskey and soda. Finishing his drink he put the glass down on a wicker table painted green. Alice rushed forward and with her handkerchief wiped the wet spot the glass had made. Father looked speculatively at her, then he thanked her. "But I don't believe I could hurt this table, Alice."

"All the same," Charlotte offered, "Father does make a ring every now and then on good furniture. Did your father do that, Alice?"

Alice looked down. "My father," she said, "did not drink."

Looking distinctly nettled Father made no remark.

"Alice is very considerate," observed Mother. She seem amused.

We spent at least part of every day in the wigwam whose charm had not yet flagged. Then one day we had The Idea. We would spend the night in the wigwam. All four of us. We planned it out in detail. Flo was to be invited for a spend-the-night party. When everyone had gone to bed we would steal down to the wigwam and sleep there. Early next morning we would come back to our beds and no one the wiser. It seemed a captivating idea. We would, of course, have a mid-night supper. To this end we began to hoard food. "Such hongry chil'ren," was Haskell's caustic comment. I felt he knew something was afoot.

We could talk of nothing but our adventure in spending the night in the wigwam. Alice was at first lukewarm. However, her thought of others this time was paying off to our advantage; she agreed to come. Fortune aided us, for the night of the spend-the-night party a friend of Father's came to make a fourth at bridge. Before she went downstairs Cousin Ella came in to kiss Alice goodnight. We would all be asleep when they finished their game, she said.

It was all arranged that Charlotte and Flo would be together in Charoltte's room and Alice and I in mine. There was a connecting door and we all gathered in my room to prepare. We each took a blanket; Alice wanted to take a pillow too, but we dissuaded her. She did go back to the room she shared with her mother to fetch her coat. She might catch cold, she said.

Lights in the servants' quarters showed we had nothing to fear at this time from Haskell and Sarah. Armed with our bundles we crept down the back stairs, through the kitchen and out to the verandah. No Indians could have moved more stealthily down the side steps and out onto the lawn.

It was a fresh, windless summer night and the scent of the flowers in the garden was delightful. We could see the lawns stretching away, their usual greenness now black and unreal. The trees looked huge and a little menacing. There was no moon, but brilliant starlight. After a bit we could see better and we wended our silent way down to the wigwam. In the daytime it was if anything too near; now it seemed a tremendous distance away. As we crept along an owl hooted and Alice clutched me and started to whimper.

"That's just an old owl, Alice." But I was glad to hold on to someone.

Now we were embarked upon this adventure it seemed different from the way we had planned it. Charlotte and Flo, for all their superior years, were clasped to each other in a purposeful grip. Such closeness hindered rather than helped progress. At times one would lose her footing and fall, bringing down her companion with her.

At last we reached the protection of the wigwam. It was black as the inside of a pocket. We stumbled in, all a bit cross. Even Alice's charity seemed to have deserted her. She was quite snappy when Charlotte accidentally sat down upon her. Finally we got ourselves sorted out with our blankets. The food was put to one side to await the hour of midnight and we sat there expectantly. We were spending the night in the wigwam!

Through the opening Satulah mountain loomed enormous and purple in the starlight. Morning seemed far away.

Some mountaineers went by on the road beyond the fence. We held our breath. Suppose they knew we were there. Suppose they were drunk. Haskell had been eloquent on the subject of "mountain dew." It was terrible stuff. Nobody in Charleston would touch it. It was an anxious time. After they had gone we relaxed and I began to think of food.

"But it's not midnight," Charlotte objected. "We are going to have a *midnight* feast."

"How do you know it's not midnight?" I countered. It seemed like midnight to me.

"Let's tell stories," Flo proposed. "Only not scary ones."

We took turns in telling stories and at

anything remotely touching upon the fearful and raconteur was adminished, "Not scary. Don't talk about anything scary."

It was in this atmosphere that we heard a rushing through the grass, then a snuffling and a hairy object hurled itself upon me . . . Flora. Alice screamed. Aghast at the possible consequences we peered out into the night. No one came. Flora settled herself against our backs and we were glad of her presence.

"It must be midnight now," I pleaded. Charlotte agreed. We would eat our midnight feast.

It was too dark to see what we were eating so instead of beginning with sandwiches as we had meant we began with chocolate cake. But I did not care; nor did Flora who licked a bit of everything impartially. While we were thus engaged, Alice, who was at the wigwam opening, reported the bridge guest was leaving. We could see his lantern bobbing along the path. Feeling very remote we watched him go in the light.

"We had better go to bed," said Charlotte with the prerogative of the eldest. We wrapped ourselves in our blankets and lay down. We had thought the ground in the wigwam was level, but now we found hollows and ridges. We wriggled around, and with Flora close to my side, I waited for sleep. It did not come, nor, to judge from the sighs and scrunchings, did it come to the others. But we lay there and waited.

If sleep was not coming something else was. We could hear quick feet coming down the verandah steps. Then, in the light of a lantern, we saw figures hurrying across the lawn. Four heads stuck out of the wigwam door with Flora's making the fifth. We had not long to wait. Mother, Father and Cousin Ella descended upon us.

Father seemed more amused than angry. We must have had a nice time in the wigwam but perhaps it would be better to come back and spend the rest of the night in our own beds. With a thankfulness we did not mention we agreed. We picked up our blankets and, Cousin Ella carrying Alice's, we went back to the house.

"How did you know we were there?" I asked.

"Alice," Mother began. But she was laughing so hard she could not finish.

Father laughed too, but all he said was, "Alice is a thoughtful child."

L.P.

Next morning after a late breakfast it seemed desirable that Flo should spend the day. We started out to acquire permission and met Miss Jones on the road in front of our gate. She was enroute to the village to pick up the mail. She seemed a bit uncertain if Flo should prolong her visit. "But John is in Flat Rock," Flo retorted. Miss Jones affected not to see the logic of this. It was perfectly clear to us. Somewhat reluctantly she agreed and we went to the wigwam.

It was pleasant there. The ferns were scented and the birds that sang and called seemed infinitely superior to the dismal hooter of last night.

Our conversation revolved about the mystery of our whereabouts being discovered. We thought we had been clever in covering our tracks. Amid the chatter I noticed Alice was silent, her usual air of competence missing. Could it be that Alice looked guilty.

"Have you any idea how they found out, Alice?"

"Well, I knew if Mother looked in and found me gone she would worry. So when I went to get my coat I put a note in her Bible saying where we were. She reads her Bible *every* night." There was a note of complacency in her voice.

We gazed at her dumfounded. There was no denying, however, we were gazing upon our rescuer. Even the bright morning sun could not altogether dispel the shaken feeling of last night. With leniency born of gratitude we forbear to blame. Looking at her in approval I murmured, "Father said you were a thoughtful child."

# *The Picnic*

It was not a nice summer. Flo and John were in England and Charlotte had outgrown me. To make matters worse we were not even at Islington. Mother had not been well so Father, remembering the many guests that place attracted, had taken Chestnut Lodge.

It was a much smaller house and there was only one room for servants. "If only we could unite Haskell and Sarah in the bonds of holy matrimony," Father sighed. "But if Haskell has eluded the married state this far . . ."

"Sarah isn't married either."

"Oh, but she *is*."

"Don't contradict your mother."

"I know, but there's Paul. You *have* to be married to have a child!" Paul was a gangling boy of about fifteen who, in Savannah, came round afternoons to bring in coal for the fireplaces. In winter there were roaring fires in each room. But not in the halls. The halls were large and in our cold spells which did not occur often... Oh life was warm and good in those days ... one would emerge from the warmth of one's bedroom and, shutting it all behind, dash at full speed down the long and winding drafty stairs. Then there was an enormous and freezing hall to traverse. Arrived at the living room door one opened it and rushed across the large room to the fire. Here one toasted oneself on both sides. Delicious. And Paul, bringer of these lovely chunks of soft coal, was undoubtedly Sarah's son.

Sarah, this summer, was staying with the relations who had made her almost miss the train in Augusta. So we had a rosy-cheeked mountain girl called Callie. She lived three miles down the Cashiers Valley road and walked to work. She did get a ride home in her father's oxcart.

There were two oxen, which with my partiality for animals, I admired. But I could not feel toward them anything like the abundance of my regard for horses. The oxen were driven by means of a single rope and the driver calling "Gee" and "Haw." It was interesting to me because it was all so different and I used to like to ride part of the way home with Callie. Most of the ride

I had to combat Mr. Roger's advice that "Effen you un don't git out now you un will have a long walk home."

It always was a long walk home. In the crowding years since that time I have often thought of that long walk home. Things can be wonderful at the time . . but there is usually a pay-off. There's usually that long walk back.

However, I must have felt it worth while. The cart smelled of the pine logs that Mr. Rogers sold for open fires so often necessary at Highlands. Everyone "took" from him. Flora Macdonald would start out with us but she soon got bored by the leisurely pace and jumped out. She disregarded oxen to the point where she disdained to bark at them as she always did the horses.

I still rode Magnolia every day and Charlotte rode Moonlight . . . but not with me. She had a close friend from Savannah, Sara Barrow, and the two of them rode together. It was, alas, a case of two's company, three's a crowd. I was told with elder sisterly firmness I was "not to tag along." Not unduly sensitive I would rush and saddle Magnolia and go after them . . . only to be told to "ride somewhere else."

So I often hung about the inadequate stable at Chestnut Lodge.

There was, however, a tennis court and the young people we knew often came and played ... especially the Eskrigge children. They were from New Orleans and Mr. Eskrigge was a cotton factor like Father. And, like Mother, he was English. The Eskrigges all played good tennis and Mother urged me to play more often. "Horses are all very well, Edith," she'd say, "and I'm glad you ride so well. But I don't want you to grow up to be one of those tough looking females who seem on a horse when they're not." That appeared a remarkable achievement to me, and furthermore, an enviable one. To be always on horseback must be heaven.

But I did play tennis and improved my game to the point where the racquet didn't feel as though it were as big as I was. Father liked tennis and it was fun to have him offer to take me on, even if I lost in straight sets every time.

I was moping about the kitchen one afternoon when Charlotte and Sara had left me behind again and was complaining to Haskell that Magnolia felt as badly about it as I did.

He made that clucking sound with his teeth. "You and *horse!* You t'ink you is a horse."

And I remembered the time when Flo had been a horse. It had been much nicer then. It was long ago, when John and I were very little, and we also at Chestnut Lodge, right next to Highfields. Flo and John and I were together all the time. We had a small wagon that had shafts on either side, maybe intended for a St. Bernard. John and I loved the wagon and somehow persuaded Flo to be our horse. She'd hold on to the shafts and pull us around up and down the uneven paths and even into the rhodendron thickets laughing and spoiling us. We even had a whip, a piece of rope tied to a stick which we thought was a fine buggy-whip and calling "giddy-up, giddy-up," we'd wield it mightily. We must have been rather feeble and merely hit her dress, for Flo merely laughed. She was a sunny good natured child. Thinking about it now, I was a little ashamed; that was outrageous of us. Now I missed them both, and envied them their wonderful ocean voyage and the trip to London. They were going to see the tennis at Wimbledon; and furthermore, visit their Aunt Janie in Rome.

"Gawd knows when you going to act like a lady!" Haskell recalled me to my unhappy situation.

"Never if it means dressing up each afternoon and either having a lot of people to tea or going to somebody's house for tea."

One morning Father told me I was to go and stay for a few days at Faraway. Charlotte would be at Sara Barrow's. Mrs. Dougal, one of the wonderful women at Faraway, was very fond of children and at least once a summer Charlotte and I went for a visit. The thought of being at Faraway with its enormous barn and all the horses was exciting. Faraway, excepting for the Eskrigge place, was the last house before the top of Satulah mountain and had really a breathtaking view. From its front verandah one could look out into four states, North Carolina, South Carolina, Georgia and Tennessee. Even I felt its rarified atmosphere over and above the lure of its well filled barn. Certainly the house was beaitiful; but it was the distinction of its occupants that gave it its unique quality. They were three women who had lived wonderful lives and had traveled together all over Europe and the British Isles collecting beautiful things.

Mrs. Dougal was the widow of a Scot. Very charming, it was probably to her that the menage owed its unusual quality. She presided at its entertainments and she kept the house. Aided, of course, by "Aunty Carolyn" who was not fascinating like Mrs. Dougal or famous like Dr. Lapham. She was just "Aunty Carolyn" and probably the bedrock of all things.

Dr. Lapham was a woman doctor, a tuberculosis specialist and the person who introduced the pneumothorax treatment to this country. What I admired her for was her horsemanship. She rode to her sanitorium and to visit her other patients. She wore a divided skirt and in wet weather had a perfect waterproof coat and cape. Her horse was named Joe and was a thoroughbred. She had, herself, gone to Kentucky to select him. I admired her tremendously; but I was more than a little in awe of her.

Crunkleton drove me up and Mrs. Dougal greeted me and showed me my room. She had put several of the Beatrix Potter books on the shelf by my bed. She talked to me as if we were of an age. I thought her enchanting . . . which she was.

She was pretty and she smelled of a faint but lovely scent. Ethel, the maid, came in and unpacked my suitcase and showed me where she put things. She also told me to wash my hands and come down straight away as lunch was almost ready.

The rest of the staff included Addie, the cook. She was a tall, gaunt, black-haired mountaineer who, under the tutelage of Aunty Carolyn, made the most delectable English muffins. I remember Charlotto's mortification when we were there together and I consumed what she considered too many. But Mrs. Dougal took up for me. I exercised a lot, she said.

Then there was Love Henry the groom and stableman. It never struck me until years later that Love was an odd name for a groom. If, like Aunt Florence, they had called their servants by their last names it might have been less startling to strangers.

That afternoon "The Faraway Ladies," as they were called, went, all except Dr. Lapham who had to be at her sanitorium, to a party. Ethel gave me tea and took me for a little walk along the road that led to Satulah mountain. Then I read some of "The Tailor of Gloucester" and by that time the ladies had returned. I flew out to the barn to watch Love unhitch the two carriage horses, Dandy and Button. He put me up on each of their backs as he led them to the watering trough.

Inside the barn was wonderful. There were the carriage, a smart dog-cart, a buggy Dr. Lapham used when it rained in torrents, and all the wonderful smells of hay, harness oil and leather; to say nothing of the good smell of horses.

And the beautiful loose boxes. It was a huge affair, this barn, and when automobiles finally took the place of carriages it was a pretty tricky thing to go up Faraway's precipitous driveway, disgorge passengers at the front steps, then drive down a slope at one side of the barn, negotiate a sharp turn into the barn, go through it and emerge headed in the right direction for that breath-taking downward driveway. But that trouble was a long way off.

Next day I tried to play with the dog, Louis d'Or. He was a Boston bull of distinctive pedigree. But he did not care for children. The ladies had gone out and I went to the kitchen and asked Addie if she would give me something for a tea. I thought it would be fun to take Louis d'Or for a picnic.

Addie looked disapproving. "How fer you aim to go. Not to the top of the mountain?"

"Oh, no. Not even as far as the Eskrigges'."

Just by Faraway the road was fairly flat. Then after you left its spring house by the side of the further gate it began to climb. I trudged along until I thought I had really gone somewhere; then I saw a bit of clearing on the valley side of the road and stopped here. Louis d'Or seemed to have had enough of my society by this time. I think he resented my too frequent pattings, (you couldn't be too affectionate with Flora Macdonald) and seizing the moment, made good his escape.

Determined not to be done out of my picnic I sat down on the edge of the road and spread out Addie's offerings. They were biscuits with crisp, if cold, bacon and chocolate cake. I had just started on a biscuit when a man emerged from the thick brush. He was gaunt, unkempt and dark visaged. He was panting as though he had come straight up the mountain side. Then he said, eyeing my picnic, "Kin I have one of them biscuits?"

I nodded and held one out. He grabbed it and ate it just like Flora Macdonald sometimes did her food. He must have been real hungry. He looked unhappy as well as hungry.

"Here," I said, "take the lot. I'm not really hungry; I just took Louis d'Or for a picnic. But he left me."

"Take 'em *all?*"

I nodded and he grabbed up the box as if I might change my mind. He wolfed down another biscuit, then said, "Kin I git to the Dillard road 'thout goin' to the town?"

"Oh, it isn't a town; it's just a little village. But you can get to the Dillard road without going there. There's a trail just along the road and you take it and go down by the Elliott's tennis court and then when you hit a road you turn to the left and soon you'll be at the Dillard road. I'll go along with you to the trail."

I got up and we walked along. I looked up at the house as we passed Faraway but did not see anyone. In a few minutes we arrived at the trail. "You just go in here. It's pretty steep. But soon you'll come to the road." Still clutching the box of food he hastily left me. Suddenly I realized he'd never even thanked me. Somehow I didn't mind; I felt something was wrong.

I don't know why, but back at Faraway I did not say anything about the picnic or meeting the man. Nor did I reproach Louis d'Or for leaving me. I realized that a little of me went a long way with him.

Next morning after breakfast I was in the huge, good-smelling kitchen talking to Addie. On the kitchen table was a copy of the Franklin newspaper. I picked it up to see what the large picture was. Over huge headlines was a photograph of a man. And those headlines:

ESCAPED MURDERER. I looked more closely. It was! It was my friend of yesterday whom I had led along the road and shown where the path was.

And Addie was saying to Ethel, the impeccable maid, "The sheriff had every road blocked; he must have had friends helping him!" She was indignant.

A murderer? That poor hungry man? I edged closer to the paper but I didn't say anything.

"A bushwhacker from back over there in Jackson County," said Addie, "moonshining and carrying on, think they can take the law into their own hands. I hope they catch him."

I still didn't say anything and looked again at the picture. I hoped he found the Dillard road and caught a train there. Ever for the underdog – a trait I got from Mother, I hoped they'd never find him.

Then I thought of John. If only John were here so I could boast of meeting a murderer. I would write him; but it would not be the same. Anyhow, Miss Jones might hear of it and something told me Miss Jones would not think well of this encounter. But John would have been impressed. I sighed. It was dull without John.

But it proved not to be. Next morning I was told that now I could go home. And what did I think I would find there? A baby brother!

Suddenly it seemed a very nice summer. When I got home I found that Sarah had arrived, comfortable beautiful Sarah. Solving the problem of sleeping quarters turned out to be very simple – she would sleep in the room with the baby and take care of Mother and the baby and let Haskell continue to run both the pantry and the kitchen. Callie would continue as the maid. The baby was very little and pink and I wondered if by next year he would be big enough so that John and I could put him up on a pony. A brother!

# *Aunt Janie*

The next year we were back at Islington and everything was glorious and John and I were practicing for the gymkhana. We would enter as many classes as we could. "Don't you ride any more just for the fun of it?" Father frowned. "It's quite all right to do your best in the show; but all this looks too anxious to win . . ."

"Nonsense," interrupted Mother, "They just want to do well."

One morning, however, another interest came into my life, Aunt Janie. I must have heard of her dozens of times, but not with any attention.

We were at breakfast. The baby was allowed to come and sat in his high chair wielding a spoon. One of the great-aunts was there, the one who made such nice jam. A wasp attempted suicide through the hole where the ladle came out of the marmalade pot and with my egg spoon I held him captive. Charlotte, who had attained the terrible humaneness of the early 'teens, was outraged. The great aunt kept saying "Watch out!" Words might have ensued had it not been for Haskell. For all the inscrutability of his black face Haskell was aware of every nuance in the atmosphere about him. He removed the marmalade pot and at the window assisted the wasp to escape. The marmalade returned to the table, we all went on eating it with undiminished appetite. As a family we were not squeamish.

Pouring Father's coffee Mother said, "She must have been very beautiful . . . Hampton's sister, I mean, Janie."

"I suppose so. I guess she was a good-looking girl. Looks run in the Perry family. She left Charleston, of course, when she married." Father paused, considering. "I never saw her to remember. Everyone acknowledged it was love on both sides. Janie hadn't a bean."

"It was sad, his dying."

"Oh well," Father said philosophically, "he was old. If it comes to that, Janie must be old now, nearly seventy."

There was an interested silence during which we watched Haskell deal with more wasps hurrying in at the window.

"'His cousins and his aunts,'" Father

sang somewhat sketchily in tune.

"Oh," said Mother, "that reminds me. I told them they could have the musicale here at Islington."

"Good heavens, Edith. Not *again*! Florence and her Highlands Improvement Society?" he inquired with wrath.

Mother looked a bit sheepish. "Florence is taking care of the young man who is coming from Atlanta to sing."

"One of those prune juice and tea occasions!"

"Fruit punch. You know I never serve prune juice, George."

"Miss Lee at school," said Charlotte, "told us prune juice is good for you. It was in physiology class. She said that every morning on an empty stomach . . ."

"Spare us the precepts of Miss Lee," said Father mournfully. Breakfast with the children was a practice of summer. He found it one of its trials.

At that point the baby tipped over his cereal, gurgling at his cleverness. Haskell left the room.

"Janie di Litta will help with the musicale. From what Hampton says she and Forence must be very alike. It will be interesting to see Greek meet Greek," said

Mother ignoring the fact that Sarah had come, cleaned up the cereal and removed her young charge.

I hit my second boiled egg a blow on the crown. "Aunt Janie," I observed, "is not Greek. She's Italian."

"Edith," said Mother, "you are not to call the Duchess di Litta 'Aunt Janie.'"

I went on imperturbably eating my egg. "Why not? Flo and John do."

"Of course. She *is* their aunt."

"Well," said Charlotte logically, "if she is Uncle Hampton's sister why isn't she an aunt?"

Mother sighed. "Uncle Hampton and Aunt Florence and Flo and John . . ."

"Sounds like your prayers," interrupted Charlotte. "You know, at the end when you start the 'God bless' list."

Mother looked patient. "Yes, well, Uncle Hampton and . . . and the rest of Highfield," she said with a quick look at Charlotte, "are not *real* relations. I knew Aunt Florence in England when we were girls and Father . . . before he moved to Savannah, of course . . . grew up in Charleston with Uncle Hampton. We love them as much as if they were family; but they aren't," she concluded in an exhausted voice.

"Nor is Aunt Janie," I added helpfully.

Looking baffled Mother said, "If you have finished your breakfast you may get down."

Haskell caught my eye. "John waiting for you with the horses."

"We're going to practice," I explained.

Father looked annoyed, but Mother said, "Now run along and make some mistakes ... which no doubt you will ..."

As I walked down the front steps John rose from the mounting block and handed me my reins. "You sure eat a lot of breakfast. I've been waiting for hours. Your stirrups are too short for show riding."

I climbed down, lowered them a hole. I mounted, gathered the double reins with the snaffle bit in both hands, put my knuckles together, held my little fingers loosely so as not to injure Magnolia's mouth, remembered to lower my heels, keep my elbows at my sides. We rode off down the driveway, my back erect. Flora Macdonald came bounding out of the tall grass, barking furiously. Moonlight danced a bit, then reared. John laughed fondly, as if she had done something clever.

Actually, of course Moonlight was Charlotte's horse, as Magnolia was mine. Mother had named them half in affection for her adopted South and half in derision of the lush Southern novels so popular just before World War I. Moonlight, skittish even with John, was impossible with any one else, except Charlotte with whom she was docility itself. But Charlotte did not care for show riding. So John rode for her in the pair entries so favored by the local show. Other events he took on his own rangy First Over.

"We'll practice for the three-gaited round Harbison's Lake," John announced with his usual authority. "Then we'll do the jumping in the paddock."

"Father says we practice too much."

"Rot. Hold your hands lower."

Round Harbison's Lake the road went through the pines and hemlocks, flat all the way, and I would mutter, "This is the forest primeval. The murmuring pines and hemlocks." It went in perfect rhythm with the slow canter we were practicing. Today, however, I was thinking of Aunt Janie. The only duchess I knew anything about was the one in *Alice*, and I pictured Aunt Janie in a headdress like Tenniel's drawing. Only subconsciously did I think of the pig.

I turned to John. "Now canter with the off-fore leading. OFF FORE, fool," he shouted.

"Fool, yourself!" Indignantly hitting Maggie with my heel, we left John behind. In equal wrath he pursued us. Just as we rounded the corner at full gallop a tent loomed whitely on the lake side. Moonlight braced herself to a stop, then jumped sideways across the road. John shot off and lay supine amid the hemlock needles. I waited for him to get up but he did not stir. Then I saw two men emerge from the tent. One was young and slim and the other was short and fat. They rushed over to John, the fat one bewailing the while.

"I think he's only knocked out, Izzy," the young man said. "He did not hit a tree when he fell off."

John opened his eyes. "I did *not* fall off. Moonlight jumped from under me. It's quite a different thing."

"Is that a fact?" the fat man asked with interest. "Well," he said after some reflection, "the result seems to be the same."

"I think," said the young man, "you'd better come in the tent and lie down, sonny."

John did not care about being called "sonny," I knew, but he was fond of tents. I tied the horses and we went inside.

"Why are you here . . . in a tent?" The accusing note in John's voice changed to envy as he gazed around.

The fat man spread out his hands in an expressive gesture. "Mr. Renaldo, here, has been ill. Pneumonia," he said direly . . . "So he had to have rest and mountain air. New York in the spring was too much for him after Italy."

"Italy?" John broke in. He had been there once visiting Aunt Janie. "Are you an Italian, Mr. Renaldo?"

"No. I've been living in Italy . . . singing. But I'm an American. My name is Reynolds . . . Alan Reynolds. I just sing under the name of Reynaldo. And this is Mr. Lipinsky who is my manager and my friend."

"Mr. Renaldo," said Mr. Lipinsky, "has been at La Scala."

"What is La Scala?" asked John.

"It's where they have the best opera in the world. Opera – singing, dancing, acting. The works!" Mr. Lipinsky waved his hands.

"The works," mused John with a worldly air, "of course."

"Have you ever seen a play?"

"Peter Pan?" I asked.

"Well, yes," said Mr. Reynaldo.

"I've seen a proper play," said John, "And I was in one at school. It was all about Columbus asking Queen Isabella if he could discover America. It was in Italy, too . . . the play was. But we said it in English."

"Is that a fact?" Mr. Lipinsky said.

"No, Spain," I said.

"Yes," John replied simply.

"Were you Columbus?" Mr. Reynaldo asked kindly.

"No, but I stood right by him. I was a courtier. One of the boys was Queen Isabella. He did look silly. Afterwards we called him Bella and he didn't like it. I was in a play, too," John went on, always stimulated by an audience, "at least it was speeches from a play."

"What did you say?"

John stood up. "My name is Norval; on the Grampion Hills I feed my father's flocks." He stopped, momentarily stymied.

"Norval, eh?" said Mr. Lipinsky. "Now what do you know about that?"

"Just in the poem. My real name is John Hampton Crop Perry. Shall I go on?"

"Yes, do," complied his audience and John beginning again, raced to a breathless conclusion.

Both Mr. Reynolds and Mr. Lipinsky appeared impressed. "You can really roll it out and that's a fact," Mr. Lipinsky observed.

I thought John had showed off enough. "I think we had better go," I said.

"Oh, don't go," urged Mr. Lipinsky.

But I turned suddenly shy. Besides there was Miss Jones. Never one to probe the future, John was inclined to stay.

"Well," Mr. Reynolds said, "come again some other time."

"We ride round the lake, practicing."

"Practicing for the horse show," John explained.

"Really? You mean a horse show like the Garden?" Mr. Lipinsky exclaimed.

We had no idea what he meant. We could not imagine anyone letting horses ruin their garden or turf, so we said nothing.

"There now, Izzy, you'll like this place. Horse shows like Madison Square Garden.

"Oh, Madison Square Garden—" we

said.

"I might ride myself. Not in the show, of course," said Mr. Reynolds.

I stood up. "Goodbye, Mr. Reynolds," I said primly.

"Renaldo," said John who seemed to have allied himself with Mr. Lipinsky.

The day on which Aunt Janie was to arrive was filled with emotion. As we were finishing lunch Aunt Florence came into the room. "Disaster, disaster," she cried, brandishing a telegram which she flung dramatically at Father who had risen to greet her.

"Has something happened to Janie?" Mother asked quickly.

"*Janie?*" exclaimed Aunt Florence irritably, "Of course not . . . it's the singer. The fool has broken his leg."

"Intentionally?" Father appeared interested. "It seems extreme."

"*George!*" Mother said. "Perhaps a garden party instead," she went on soothingly.

Aunt Florence was not to be soothed. A gleam had entered her eye. "George," she said, "you sing quite well. And there is that young man in the village, Will Bacon. I know he is not exactly right in the head," she added quickly. "But they say he has a wonderful voice."

"Will Bacon," I put in, "has a banjo. P'raps he could bring the banjo and . . . "

"Florence," said Father sternly. "I will *not* sing, with or without the village idiot. Please realize that."

"We might have tableaux," Aunt Florence acknowledged defeat. "Or the children might do recitations."

"People expect to be fleeced for charity," Father said, "But to take money to hear your son being Norval on the Grampion Hills is larceny after trust."

"P'raps."

"We've got another week," Mother said hopefully. "Something will turn up." Aunt Florence appeared not to share this Micawberish hope. She went off gloomily. "Janie comes this afternoon," she called back as if reciting a final blow of fate.

That afternoon Charlotte and I decided to go to Highfield for tea; we had every intention of being there when Aunt Janie arrived. Uncle Hampton had gone to meet her at Seneca where the New York train stopped. They were to drive back in Dillon's Ford and that alone would be worth seeing. Seneca was even further away than Walhalla over rugged mountain roads, the radiator cap of Dillon's Ford never stayed the course and we knew by the time the car came up the Highfield driveway it would be a Chimaera blowing steam and smoke in a very satisfactory way.

Nor were we disappointed. At about five o'clock we heard a terrific chugging and rushed outside. Amid smoke and steam and the sound of machine gun firing Dillon's Ford came to rest on the turnabout known as The Rock. Uncle Hampton stepped out in an exhausted way and paused to help a tall, slender lady descend.

"Ah, Florence. How lovely to see you," she said and I knew it was Aunt Janie. Her voice was like music, the remnant of a Charleston accent part of its charm.

Aunt Florence kissed her warmly on the cheek. "Now do come inside, you must be tired."

"But I must greet the children," cried Aunt Janie. She turned to us. She had on a little traveling hat that only partially covered her head. Her hair was neither dark nor grey, but waved in thick strands of black and white. Dappled like a pony, I thought. She had an elegant little nose, enormous dark eyes and a smile that was at once wistful and engaging. She who had been so loved and admired seemed to be asking: "Oh do be pleased with me!" I looked at her and loved her instantly and always.

Aunt Florence brought Flo forward. "She has your beauty and Hampton's phlegm."

John advanced with outheld hand, on his face the blankness it assumed in public. Aunt Janie shook hands. To his obvious relief she did not offer to kiss him.

Aunt Florence touched Charlotte who stepped forward and made her graceful curtsy. Aunt Janie seemed charmed. "Blue eyes and black lashes. Irish, isn't she?"

"George Inglesby's girl. Not a drop of Irish blood in her," Aunt Florence said with finality.

"Pity," Aunt Janie smiled her mischievous smile, "Or is it? All the charm in the world . . . the Irish. But no stability. Do you remember Rory O'Rourke who used to come over to England for the polo?"

"No," answered Aunt Florence, "but I remember hearing of him."

"Of course," said Aunt Janie tragically, "You are too young. I forget. No one is as old as I; no one can remember my life now." She suddenly assumed an air of great fragility and age. I was stricken on her behalf; but Uncle Hampton merely looked annoyed. "Now Janie," he said, "That is nonsense."

Charlotte shoved me forward and Aunt Janie ceased to be old and tragic. I kicked the ground behind me in my version of a curtsy and basked in the glory of her smile.

"Another of George Inglesby's girls," said Aunt Florence. "Half English half Charlestonian, not Irish at all."

Aunt Janie looked down at me. "Not Angles but angels."

"Not too much of an angel," said Aunt Florence. "She and John are always up to some mischief."

"What? That proper little Englishman!"

John's eyes, studiedly blank, now assumed a speculative gleam. He looked at Aunt Janie with interest.

Uncle Hampton stiffened. "He is American. After all, Janie, he is my son."

"Half yours and wholly Florence's."

At this point Dillon, a fascinated spectator, joined the party. "Did you get the telegram all right, Mrs. Perry? When I took it over the 'phone I saw you across the street at the Post Office so I sent it over by Will Bacon. He ain't very knowing."

"I got it. But you shouldn't be so casual with telegrams."

"It was on account of my mare, Goldy. She was dropping her foal and I was so busy I let Will take the wire."

Aunt Florence's face lit up with interest. "Colt or filly?" she inquired.

"Colt," said Dillon with pride.

"Good," John put in. "Don't touch him 'til I get back from school next summer. I'll help you. If I were just going to be here this winter," he said regretfully, "I'd have him ready for a snaffle bit by summer."

Dillon appeared not too gratified by John's offer; it seemed the last straw to Uncle Hampton.

"Florence," he said in a tone of suffering. "This hardly seems to be the time . . . Do bring Janie into the house and give her a cup of tea."

But Aunt Janie had turned to Dillon. "You brought us up the mountain beautifully, my man," she said with a radiant smile. "Now tell me," pointing to an odd looking searchlight, "why is that contraption on the front of your car?"

"Janie," said Uncle Hampton in

suppressed fury, "if you don't want your tea, *I* do. And what is more, I need it."

The Highfield contingent started indoors, Charlotte and I bidding them goodbye. "You-uns want a ride to your gate?" Dillon inquired. This was fun. We departed in a series of explosions and were at our gate before I had time to reflect that Aunt Janie had not worn a headdress.

The next day was Sunday. We liked church. To begin with we were all fond of singing and roared out the hymns and chants with vigor. And usually there was entertainment of some kind as on the day old Doctor Elliott from New Orleans came in followed, unknown to him, by his Irish setter, Reddy. Not the least upset, Doctor Elliott had merely put Reddy in the narrow aisle next to the window and there he sat through the service like the gentleman he was.

We could not hope for plums like this often; but Will Bacon who was what the villagers described as "wanting," sometimes came and sang . . . in tune and well . . . but in a manner quite different from the rest. It was the way mountaineers sang, Miss Jones explained. She herself being Welsh sang with fervor and beauty.

We sat with our Sunday, composed faces. There was a rustle and a small commotion and Aunt Florence came down the aisle followed by Aunt Janie and Uncle Hampton. All eyes were glued upon them. At the Perry pew Aunt Florence paused and motioned Aunt Janie to proceed her. Aunt Janie smiled her enchanting smile and motioned Aunt Florence onward. For two cents, to judge by his expression, Uncle Hampton could have knocked their heads together. Then Aunt Janie made a low genuflect and went into the pew. Evidently the English church in Rome was High. I could see Miss Jones, Low Church and firm about it, bristle. As for us, we were delighted.

But wonders had not ceased. John turned and gave me a look like a nudge. Entering the pew across from us were Mr. Reynolds and Mr. Lipinsky. Mr. Reynolds turned and gave me a slight smile. I felt like one publicly receiving citation. John was gazing fixedly at this friend Mr. Lipinsky. That gentleman, his head slowly revolving, was drinking in surroundings no doubt strange. He must have seen John; for surely, I reflected, there was no one else in the Highfield pew to whom he would give that slow and altogether delightful wink. Church was wonderful today.

Later, at the Rector's "And now with . . ." we all leaped to our feet in grateful release. Mother whispered that it was her turn to stay and remove the altar flowers. We were to wait outside.

John and I waited for Mr. Reynolds and Mr. Lipinsky. Aunt Janie was holding court in the church porch. We saw that Mr. Reynolds was staring at her. "But it could not be. Who *is* that lady?" he murmured.

"Oh that?" said John, "that's my Aunt Janie."

"Aunt Janie?" repeated Mr. Reynolds slowly, "surely, that is the Duchess di Litta."

"Well, she is a duchess. Are you going to rent a horse and ride, Mr. Reynolds? Mr. Lipinsky could . . ."

But Mr. Reynolds had gone forward. He waited 'til Aunt Janie was disengaged then he went up to her. To our astonishment she smiled her famous smile and said, "Alan." To our even greater wonder Mr. Reynolds took her hand and gracefully kissed it. Before he could speak she said, "How splendid to find you here. I recognized you when you sang. I turned to see where that glorious sound was coming from . . . and there you were. It was in the *Te Deum Laudamus*, you know. I whispered to Florence. "It really is *Te Deum Laudamus* . . . for there you have your man."

Mr. Reynolds laughed, "And how is that?" But Mr. Lipinsky was making his presence felt. Mr. Reynolds looked at Aunt Janie. "May I present my manager? Mr. Isadore Lipinsky. The Duchess di Litta."

Mr. Lipinsky gave Aunt Janie's hand a hearty shake and looked bewildered as she said, "Now what could be better? You can look after it all for us."

Mr. Reynolds laughed. "How like Rome this seems."

"Well, it isn't Rome and you are a godsend."

She called to Aunt Florence, "Do come over here, Florence, and we'll get it all arranged now. Florence," she went on, "this is Alan Reynolds whom I knew in Rome. He sings in opera. I do think he will be better than that wretched man who broke his leg. So you can be grateful for the poor thing's misfortune."

Aunt Florence fixed Mr. Reynolds with a conqueror's glance. "Then that is settled," she said, "And Edith will play the accompaniments."

She looked about for Mother who was still absent attending to her altar duties.

"All quite perfect," said Aunt Janie.

"Now wait a minute, Duchess," said Mr. Lipinsky. "This fellow is under contract."

"Nonsense," said Aunt Janie. "It will not in the least conflict. We are not going to *pay* him anything."

Mr. Reynolds laughed and laughed. "I don't know quite what this is all about; but I do know the Duchess always gets her way. However," he added, "I think Mr. Lipinsky had better do the accompaniments. Without pay, of course. Eh, Izzy?"

Miss Jones had lingered to speak to the Rector. She now caught us up and began herding us away. John said pompously to me, "Well, *we* knew them first. If Mr. Lipinsky hadn't met us . . . "

"Aunt Janie knew Mr. Reynolds before," I broke in in the interest of fair play. "But I do think it helped . . . his knowing us."

We did not, of course, attend the musicale. But we watched the arrival of Mr. Reynolds and Mr. Lipinsky from the upstairs verandah. John so far forgot himself as to whistle down to Mr. Lipinsky who obligingly whistled back. The concert . . . a minor event after this excitement . . . was said to be a great artistic success, and, as Aunt Florence shrewdly observed, was as great a financial one.

A few days afterwards we were playing Flinch on the piazza at Highfield. Aunt Janie, Mother and Aunt Florence were having tea in the drawing room. In search of scones and shortbread we wandered in. Aunt Janie was saying, "Edith, you scooped the Metropolitan. Alan is singing there this winter. But he sang here before a single New York critic heard him."

Laden with provender we returned to our Flinch game. John said to me, the scorn in his voice slightly diminished by shortbread, "Uncle George said we practiced too much. Well, if we had not gone round Harbison's Lake they'd never have had their ole concert. And I'd like to know what the horse show would be without practice."

I nodded in complete agreement. The importance of the horse show was not to be trifled with. After all, I thought with passion, horses are the real business of summer.

# *The Winged Chariot*

Our clothes, better yet not up to party standards, gauged to a nicety the climate of the afternoon's expedition. We were going to tea with Miss Alicia.

Mother called, "Have you proper handkerchiefs?" It was her usual parting remark.

With patience Charlotte said yes. Thirteen, she was beginning to care about such things.

"I thought Flo and John were going with you?"

"They're meeting us at the gate."

Outside our grounds they were waiting. "We must be sure," Charlotte said, "not to talk of horses. Remember, Edith . . . you and John."

"Yes, We mustn't talk of love, either."

"No," Flo agreed.

I was a solid child with a passion for horses, and although Flo was slender with her mother's vivid beauty, she had Uncle Hampton's phlegmatic mien. A more unlikely pair to discuss the tender passion would be hard to find.

Once each summer we were summoned to tea with Miss Alicia . . . a not unmixed pleasure. Miss Alicia had for us the fascination of the macabre. She was a cripple. The story as it came to our innocent ears, was spectacular enough. Miss Alicia had been a beauty and a belle. One winter morning at a house party near Charleston her horse had fallen and rolled on her, injuring her so she could never walk again. That was bad enough; worse was to follow. The young man in Charleston to whom Miss Alicia was engaged *asked to be released*. Certainly, we felt, no one could have had a more dramatic past.

Miss Alicia was old, so old that for us there were two of her . . . the girl of the romantic story and the old lady of whom we were slightly afraid. She was the last of her family and each summer came up to Highlands alone except for her maid and her butler, Manigo. Manigo, Father said, had absorbed some of the ecclesiastical flavor of Miss Alicia's father who had been a bishop. Miss Alicia was striking with her Roman head and her dark, flashing eyes; but no one had the presence of Manigo.

If Miss Alicia was immobile her voice took her out into the world. It was sweet but piercing and she could throw it almost like a ventriloquist. We would be trooping past her cottage, momentarily silenced, on tiptoe so to speak. We would not see her on the verandah deep in the shadow behind the rhododendrons; but she would call out so compellingly that, defeated, we would file obediently up her garden path.

"She always has a rug over her lap. Do you suppose she has any legs?" John asked.

"Of course." Then more uncertainly Charlotte added, "At least I suppose so."

We trudged along with awesome thoughts. Then John said, "We won't go through the village. Let's go round by Dillon's stables."

"That will just start you and Edith talking about horses and you'd be sure to forget."

I knew why John was avoiding the village. In a day or two he was going back to a school chosen by Aunt Florence for its adherence to English methods. Now he wore flannel shorts, a dark blue blazer with his school device and a little cap on which his school colors were emblazoned. He had not his usual aplomb.

Mother let us grow up in Savannah as our friends did. Not so Aunt Florence with Flo and John. They were as English as English nannies and governesses could make them. With the result . . . and well they knew it . . . that in England they were Americans and in America they were English.

Most of us at Highlands had summer homes that were on the mountain. Miss Alicia's cottage, however, was on level ground. We went round the village then down between thick clumps of rhododendron. It was a gloomy spot and we were harboring gloomy; if exciting, thoughts. John and I thought of Miss Alicia's horse: Charlotte and Flo, who were beginning to enjoy romantic ideas, concentrated on the enormity of her beau's defection.

"I wonder who he was," Flo said, "and if he still lives in Charleston? Do you suppose he's married now and we know his children?"

"Grandchildren."

"Yes. It was a long time ago."

Shocked, we digested this thought. Miss Alicia was to us not so much a person as a legend. We created in our imaginations a picture of a beautiful girl, gay and spirited . . . and doomed. Carried away, we forgot the real Alicia, the one we were going now to see. The remembrance brought us up sharply.

After a moment Charlotte said, "Oh, he wouldn't ever *marry*. He'd at least stay true to her."

"Doesn't seem very true, leaving her like that when she was hurt."

Their voices, arguing the ethics of the unknown's conduct, died away in the intensity of John's and my talk. Horses were the breath of life to us. Anything touching upon them was important. What John passionately wanted to know was whether Miss Alicia was riding side-saddle at the time of the disaster. "Tied up in all those pummels she couldn't jump free. Do you suppose we could sort of bring up the subject of side-saddles and see what she says?" Controversial by nature, discourse with John was one long academic question.

"Charlotte says we mustn't even mention horses."

"Well," he observed ominously, "I hope she has a good tea."

"Oh, she will," I said fairly, "She always does. She has marvelous things to eat."

On this encouraging note we caught up with the others at Miss Alicia's gate.

Miss Alicia had a beautiful garden; though it did not interest her. People were what she liked. People and the sound of the human voice bringing in the world outside. Manigo, however, was a gardener at heart and a butler by profession. He let himself go at Highlands and the garden was a riot of color and scent.

At the gate we had a moment of panic. We stood shadowed by a flowring bush, thinking ourselves invisible in its shade. Charlotte's hand was on the latch and each of us drew a deep breath like a diver about to submerge. Then the stillness was shattered by Miss Alicia's penetrating shout, "Come in, darlings, come in." Her voice was like a net thrown over us. Caught, we trooped up the path between Manigo's flowers . . . toward those compelling accents and that enveloping presence.

Miss Alicia was sitting in her wheelchair on the verandah. For one so stationary she was remarkably active. She swooped long arms and clasped each of us to her capacious bosom, planting a resounding kiss on our cheeks. What made this ordeal particularly frightful for me was the large mole on her chin. It had two hairs that stood out and I thought if they touched my cheek I should die. I could never decide if they had or not; but I always rubbed my face with vigor. Due to his manly sex or greater agility John escaped the embrace. His determined, outstretched hand was shaken.

"Now sit down, lovies. Sit down. We'll have tea . . . lots of sandwiches and cake," she fluted. "Now where is Manigo? Go, one of you and call him."

"I'm right here, Miss 'Licia," Manigo had a commanding voice. He stepped forward and drew the tea table near her.

Miss Alicia looked animatedly about. "Now where is Mrs. Gregory? Amy, Amy, do come; the children are here."

"I'll tell her, Miss 'Licia." Manigo went away with his measured tread.

"Mrs. Gregory has come to stay with me," Miss Alicia told us. Then she lowered her voice to a sibilant whisper. "She has a daughter who's married to an admiral. They are in England now and every time any one comes she reads this letter about how important they are." She turned her head and spoke in a conspiratorial whisper,

"Do children, if she wants to read it to you, *stop* her. I declare," she said emphatically, "I just could not hear that letter again."

"Did they see the King?" John asked. He had seen the King once.

"I suppose so. They wouldn't be satisfied with less."

"I saw the King in a parade. He was riding the most wonderful horse . . ."

*Horses*! Charlotte gave him a warning look. "Is your friend from Charleston?" she inquired in the voice of one changing the subject.

Then Mrs. Gregory came out. "This is Mrs. Gregory. She used to know your fathers."

"Who *are* their fathers?"

"Do, Amy," Miss Alicia looked annoyed, "give me a chance to tell you. I was going to. This is George Inglesby's girl," as Charlotte stepped forward, "They live in Savannah now. And this is another of his daughters," turning to me.

"Savannah?" queried Mrs. Gregory, "Whatever for? *I*," she added, "have never ceased to long for Charleston."

Miss Alicia gave her shrill laugh. "Come, Amy. You talk that way but you wouldn't miss the excitement of living with the

Admiral." She stopped abruptly, apparently realizing into what channels she was leading. "These two," she said quickly, "are Hampton Perry's children. You remember the Legare Street Perrys?"

"Oh yes. He married an English girl."

"Manigo," yelled Miss Alicia. "Manigo, do bring the tea."

"An English girl," Mrs. Gregory repeated. "My daughter is married to an Admiral and they are in England now. I have just had the most interesting letter . . ."

"Here's Manigo," said Miss Alicia, relief in her tone. "Now children," in her piercing yet swooning voice, "Let's all eat lots. Lots of sandwiches and cakes . . . and I think there is some candy."

We had every intention of eating lots; but we said nothing. Miss Alicia's expansiveness left us dumb.

"Cat's got everybody's tongue," pouring out tea. "Oh dear, Amy. You take lemon and I think Manigo forgot it."

"It's here, Miss 'Licia," said Manigo, taking offense. "Right behind the cream pitcher."

"Yes, so it is. Now children, what have you been doing with yourselves? Tell us all about it."

As almost all John and I did was ride we rightly interpreted Charlotte's look and said nothing. Flo looked stolidly ahead and Charlotte began to mutter something about tennis. But Miss Alicia did not hear. She and Mrs. Gregory had entered into an argument as to the merits of lemon versus cream in tea.

"Well," Miss Alicia apparently recollected us, "Have you had any nice parties? Charlotte is growing up so fast it soon will be time for parties and beaux, eh Charlotte?"

We gasped. Miss Alicia being up the subject of beaux. And so of love! We stared in amazement. Wasn't her heart broken after all?

"Yes," Mrs. Gregory looked interested. "Before you know it it will be time for her to come out and go to the St. Cecilia. Do you remember, Alicia, your first season with all the new clothes . . . and the beaux? It was fun. Not my *first* year," she added reflectively. "I was so timid. Of course Papa and the boys helped at the balls . . . and their friends, too. Besides," complacently, "I was a good dancer. It was the house parties and picnics I dreaded. I could not swim very well and I was scared to death of a horse."

Miss Alicia smiled.

"But you weren't afraid of anything," went on Mrs. Gregory in admiration. "You were . . . *intrepid*, Alicia."

Miss Alicia gave a dry little laugh. "P'raps too intrepid."

Mrs. Gregory's round face puckered in distress. "Oh, 'Licia, 'Licia. I am so sorry. I wouldn't have . . ."

Miss Alicia gave the plump hand a brisk smack, a gesture obviously intended as a reassuring pat. "Nonsense. Don't be upset, Amy. It was all long ago . . . the accident . . . and all."

They both stared into space plainly thinking the same thoughts. Then Miss Alicia said, "Miles Frobisher's horse had more spirit than he did. Miles had a handsome face, but that dreamy air covered dullness, really . . . not the poetic nature we thought."

The distress on Mrs. Gregory's face was succeeded by lively interest. "He grew prodigiously dull and enormously fat. You were well rid of him, 'Licia."

Suddenly the two ladies were aware of four pairs of eyes staring, round and incredulous.

"Now children," Miss Alicia's robust air returned, "make a good tea."

Our mouths too full for reassurance we sat and devoured food in greater quantities than even we liked; but we were unable to stem the flow of Miss Alicia's hospitality. Finally she seemed convinced we were replete. "I know, why don't all of you run out into the garden. She pronounced it "gyarden" in the old Charleston way, "and gather bouquets for your mothers."

The fresh mountain air was like freedom regained. But the garden did not seem the happy place we remembered. The scent of the pinks and roses was gone with the summer. Now the air was pungent with chrysanthemums and dahlias. Autumn was here and our mood autumnal. In a desultery way we gathered flowers as bidden; then with our hands full we sat on a rock under an apple tree. Usually we chattered like magpies about ourselves; nothing else interested us much and the flow of our own affairs was endless. Today we were quiet and oppressed, puzzled as to what to do. Miss Alicia did not seem to want us. She and Mrs. Gregory was deep in talk. Miss Alicia's piercing voice, broken by her reckless laughter, floated out to us. We stayed where we were; we had had enough of grown peoples' revelations.

Then Flo's ear began to ache. "Does it hurt much?" Charlotte assumed charge.

"No, not really. It hasn't begun to crackle . . . or pop. But it might. Do you think we could go home?"

We all got up with alacrity. Charlotte pondered. "I suppose so. You couldn't call this 'eating and running,' could you?"

"Oh no. We've been here *ages*."

A solemn processional made its way to the verandah. Miss Alicia looked surprised. The thought occurred to me that we might just as well have left from the garden. We had been quite forgotten. She was, however, voluble about the party breaking up. "But if Flo has an earache she must go home. Her mother will know what to do."

"Papa always blows smoke in my ear."

"Smoke?" Miss Alicia was horrified. "Warm oil is the thing. You tell your mother to drop a few drops of warm oil in your ear. That will fix it."

We began our farewell; but suddenly she stopped us. "Candy. You must have a piece of candy before you go. Manigo. Manigo, what have you done with the box of candy?

I meant you to leave it here."

Manigo materialized ... a very substantial genii. We each took a piece of candy, halting almost imperceptably to select the largest our eye could quickly discover. Flo, about to eat her's was arrested by Miss Alicia. "Maybe you'd better not eat it with an earache."

"Nonsense," said Mrs. Gregory. "It will do her good."

We all looked stricken at the thought of Flo enjoying her candy on the way home ... when we had finished ours. Perhaps Mrs. Gregory's experience was greater for she said, "All of you take two pieces and keep them to eat later."

As Manigo passed the box again Miss Alicia said, "It's very disappointing our nice party had to end so soon ... very disappointing indeed."

"Yes," observed Manigo with his episcopal air, "Yes, Miss 'Licia. We points and Gawd disappoints."

To this flight of theology even Miss Alicia had nothing to say. Seizing the moment we said goodbye and murmuring our thanks made a quick departure. As our recessional figures filed down the path Miss Alicia's voice floated after, "Warm oil," she called. "Remember, warm oil, w-a-r-m oil."

Several yards from the gate I stood still in anguish. "I have left my jacket."

"I'll go," offered John. This chivalry did not fool me. He was thinking of the candy. But it was not so tempting to me now.

"No, Edith must go. She's got to learn not to be so scatter-brained."

"I can't. I just can't go back."

"Why do you mind? It's so silly of you."

I blurted out a passionate protest in which the words "Miss Alicia's mole," could be heard.

"Oh. She won't kiss you. Nobody ever kisses you *twice*. Only when they meet you."

"Are you sure?" I had a touching belief in Charlotte's worldly knowledge.

Behind me the garden gate clicked and my anxious feet, moving so desperately, crunched the gravel revealingly. Miss Alicia and Mrs. Gregory looked up from their conversation, but Manigo came to meet me. Evidently he had found my coat; bearing it in both hands he advanced down the steps. I met him halfway. He stopped and with his pontificial air invested me with, rather than helped me into it. It was like the laying on of hands.

In an excess of gratitude I exclaimed, "Oh, thank you, Manigo. Thank you, Miss Alicia . . . and Mrs Gregory." On wings of thankfulness I flew back down the path, through the gate and caught up with the others.

The slanting sun glinted in the mica of the rocks that were scattered over the road and irradiated the puffs of dust that marked our progress. Ahead of us Satulah mountain loomed, its patches of russet and scarlet the first signal flares of autumn. There was a cold wind blowing. I buttoned my jacket.

At the mill pond bridge we automatically stopped to lean on the rail, but we were too preoccupied to spit to see the rings it made. Downward the water slid smoothly over stones to end in a still brown pool – usually the object of goosefleshed speculation. Some said it was bottomless; others held it to be floored with quicksand. John was sure if anyone fell over the rail he would slide straight down, and if he did not start swimming at once the quicksand would "get him." Today, however, there were no ghoulish flights of fancy. We did not feel the need.

The afternoon, begun with such spine-tingling anticipation, had somehow lost its savor. Charlotte appeared to be thinking long thoughts. Suddenly she burst forth. "Mrs. Gregory was wrong. It will be a long time before I'm grown up."

"I don't want to grow up," I said defiantly.

"Nor do I," said Charlotte somewhat to my surprise. "Just look what happens when you do and have to go to all those balls and things."

John laughed. "Charlotte's afraid a horse's going to fall on her."

But Charlotte shook her head. She looked as if she was thinking instead of the vast grownup world whose crushing weight she would one day feel.

For the first time we were conscious of the inexorable nudging of time at our backs . . . and the way exciting chapters of life could, by different readings, be flat and colorless.

"Did you hear Mrs. Gregory say Miss Alicia's beaux got fat and stupid?" Flo asked. "I'd always thought of him as being sort of dashing and handsome. Everybody says how beautiful *she* was." Her voice trailed away. I think we all felt disillusioned and a trifle elderly.

At our gate we stopped, preparatory to the parting of ways. Taking a long breath I said, "Well, we don't have to go to see Miss Alicia again until next summer."

And even with the wisdom that was darkening our lives next summer seemed a long, long time.

Inglesby, Edith

The Happy Highways

DATE DUE

| | | | |
|---|---|---|---|
| AUG 22 1983 | SEP 18 1984 | | |
| SEP 25 1983 | 10/23/84 | | |
| OCT 31 1983 | 11-13-80 | | |
| 11/28/83 | 11-27-81 | | |
| JAN 5 1984 | 2/27/86 | | |
| 1/24/84 | MAR 13 1986 | | |
| 2/14/84 | 3/19/87 ill OR | | |
| 3/7/84 | | | |
| APR 10 1984 | | | |
| MAY 14 1984 | | | |
| JUN 4 1984 | | | |